D0875372

Fourier Series

International Series in Pure and Applied Mathematics

William Ted Martin and E. H. Spanier, Consulting Editors

Fourier Series

Robert K. Ritt, Ritt Laboratories, Inc.

McGraw-Hill Book Company
New York St. Louis San Francisco Düsseldorf
London Mexico Panama Sydney Toronto

Fourier Series

Library of Congress Catalog Card Number 71-118801

52970

1234567890 MAMM 79876543210

This book was set in Modern by The Maple Press Company, and printed
on permanent paper and bound by The Maple Press Company.
The editors were E. M. Millman and Barry Benjamin.   Matt Martino
supervised the production.

Preface

I have attempted, in this book, to discuss in a mathematically concrete and elementary fashion the material that is essential to the understanding and intelligent application of the theories of harmonic analysis and generalized orthogonal series. For the student of mathematics I hope to have supplied the sinewy stuff which underlies the theories of topological groups and topological vector spaces and for the student of science and engineering a deeper understanding of the tools of his trade.

The table of contents is an outline of the development. Those entries in parentheses are treated as problem material. It is assumed that the reader is familiar with the concepts and techniques generally presented in courses in advanced calculus and ordinary differential equations. The functions we deal with are Riemann integrable and, for all practical purposes, piecewise continuous. The extension of the theory of integration would permit us to complete the proof of the Parseval theorem for the Fourier integral and to introduce, in a precise fashion, the Dirac delta function and its derivatives. However, to do this properly would have extended the text beyond the elementary level that I wished to maintain.

The text proper is designed primarily as a mathematical development.   It is interspersed with problem material which is of incidental
mathematical interest, but which, in fact, covers a wide variety of the
applications of this material to problems of phenomenological and statis-
tical sciences.   To minimize interruptions and at the same time to exhibit
the applications, I have supplied several appendixes:

In Appendix A, entitled Continuum Dynamics, starting with a
mathematical representation of the principle of conservation of mass, we
first obtain the equations required to describe the kinematics of continuous media.   Using explicit assumptions about the nature of the internal
forces in a continuous medium, we then find the dynamical equations,
based on the conservation of momentum.   Finally, from the conservation
of energy and the principles of thermodynamics, we extract the form of
the dynamical equations appropriate to fluid dynamics and elasticity.

In Appendix B, the physical laws of electromagnetic theory are
given axiomatically in the form of partial-differential equations.   The
Sommerfeld radiation conditions are first shown to be a consequence of
the form of a radiation field and are then shown to ensure the uniqueness
of solution of exterior-boundary-value problems.   The interior-boundary-
value problems are discussed.

In Appendix C, the classical heat equation is obtained as an account-
ing of thermal energy.   An example of its solution is given as a tribute to
the fact that it provided the original impetus to the subject of Fourier
series.

Robert K. Ritt

Contents

of functions having continuous second derivatives and which satisfy the boundary
conditions and representation of solution of inhomogeneous problem;
representation of characteristic values λ_n and characteristic functions y_n
for large n.

Introduction

THE PARTIAL-DIFFERENTIAL EQUATIONS OF MATHEMATICAL PHYSICS

The mathematical material of this book was born, developed, and reached maturity during the nineteenth and early twentieth centuries, in an environment of vigorous expansion of the physical sciences. This environment is now referred to as *classical physics*. With the advantage of hindsight, we can now pick out the essential features of this development and, in a relatively few pages, derive the mathematical representations which are used to predict and explain most of what can be predicted and explained. The word "derive" means that an initial mathematical formalism is unambiguously related to basic physical assumptions, and redundancies in the formalism are then created. These redundancies are created by a manipulation of the mathematical formalism, with the occasional introduction of modifications which correspond to assumptions that certain physical effects can be neglected. The form of the redundancies is that certain physically identifiable quantities are shown to be those solutions of certain partial-differential equations which satisfy

1

some additional conditions, known variously as *boundary* and *initial conditions.*

The general feature common to the appendices is the exhibition of partial-differential equations whose solutions are required to fulfill certain ancillary conditions. These conditions arise in a natural way from the physical situation; they serve the mathematical purpose of completely specifying the solution. Let us consider several simple illustrations.

The interior Dirichlet problem   Consider *Laplace's equation,* the partial-differential equation

$$\nabla^2 u \equiv \frac{\partial^2 u}{\partial x^2} + \frac{\partial^2 u}{\partial y^2} = 0 \tag{L}$$

If R is a region in the plane, a function $u(x,y)$ is said to be a *solution* of (L) in R if u has second partial derivatives with respect to x and y in R, and if, for each point in R, the sum of the values of these partial derivatives is zero. If R is a bounded region whose boundary Γ consists of a finite number of components, each a simple closed curve, and f is a function on Γ, the interior Dirichlet problem consists of finding a solution of $u(x,y)$ in R whose values as the point (x,y) approaches a point (x_0,y_0) on Γ have as a limit the value of f at (x_0,y_0). If certain hypotheses, which we shall not discuss here, are imposed on Γ and f, this problem is said to be *well-set:* First, there is at most one solution, and second, a solution, in fact, exists.

The proof that there is at most one solution is very simple. If u_1 and u_2 are solutions, then $u = u_1 - u_2$ is a solution of (L) in R, and u has the limiting value of zero at Γ. Since

$$\nabla \cdot u \, \nabla u = |\nabla u|^2 + u \, \nabla^2 u = |\nabla u|^2$$

if Γ permits the application of Green's theorem to the limiting value of $u \, \nabla u$, then

$$\int_R |\nabla u|^2 \, dR = 0$$

and so u is a constant in R whose limit at Γ is zero; thus $u = 0$ in R. The proof that a solution exists is not simple; however, for particular boundaries Γ, the methods we shall discuss give an explicit representation of the solution as an infinite series of functions with sufficiently well-defined properties to permit the proof that the representation is, in fact, a solution. For example, if Γ is a circle with radius a, and r and θ are the usual polar coordinates, we shall see that

$$u = \frac{1}{2\pi} \sum_{n=-\infty}^{\infty} \left(\frac{r}{a}\right)^{|n|} \left[\int_0^{2\pi} f(\theta')e^{-in\theta'} \, d\theta'\right] e^{in\theta}$$

The inhomogeneous Dirichlet problem   Here we consider *Poisson's equation*, the partial-differential equation

$$\nabla^2 u = f(x,y) \tag{P}$$

This time, $u(x,y)$ is a solution of (P) in R if u has second partial derivatives with respect to x and y in R and if for each point of R, when the numerical values of these derivatives are inserted into the left member of (P), the resulting value is the numerical value of f. The Dirichlet problem is to find a solution of (P) which has the limiting value zero as Γ is approached. It is again simple to prove that at most one solution exists, and not simple to prove that at least one solution exists. Using the methods of the text, it is possible to find an explicit representation of the solution.

The one-dimensional wave equation   Here we look for a solution $u(x,t)$ of the partial-differential wave equation

$$\frac{\partial^2 u}{\partial x^2} - \frac{1}{c^2} \frac{\partial^2 u}{\partial t^2} = 0 \tag{W}$$

in the region $0 \le x \le L$, $0 < t$, for which

$$u(0,t) = u(L,t) = 0 \qquad t > 0 \tag{B}$$

and

$$\lim_{t \to 0} u(x,t) = f(x) \qquad \lim_{t \to 0} \frac{\partial u(x,t)}{\partial t} = g(x) \tag{I}$$

To see that at most one solution exists, let u_1 and u_2 be solutions. Then $u = u_1 - u_2$ is a solution which satisfies (W), (B), and (I) for $f(x) = g(x) = 0$. Now,

$$\frac{\partial}{\partial t} \left\{ \int_0^L \left[\frac{1}{c^2} \left(\frac{\partial u}{\partial t}\right)^2 + \left(\frac{\partial u}{\partial x}\right)^2 \right] dx \right\}$$
$$= 2 \int_0^L \frac{\partial u}{\partial t} \frac{\partial^2 u}{\partial x^2} \, dx + 2 \int_0^L \frac{\partial u}{\partial x} \frac{\partial^2 u}{\partial x \, \partial t} \, dx$$

If the first of these integrals is integrated by parts, the right-hand side of this equation becomes

$$-2\left[\frac{\partial u}{\partial x}\ \frac{\partial u}{\partial t}\right]_0^L$$

But from (B),

$$\frac{\partial u(0,t)}{\partial t} = \frac{\partial u(L,t)}{\partial t} = 0$$

Therefore

$$\int_0^L\left[\frac{1}{c^2}\left(\frac{\partial u}{\partial t}\right)^2 + \left(\frac{\partial u}{\partial x}\right)^2\right]dx$$

is a constant. As $t \to 0$, $\partial u/\partial t \to 0$. Therefore this integral is

$$\lim_{t\to 0}\int_0^L\left(\frac{\partial u}{\partial x}\right)^2 dx$$

But

$$\int_0^L\left(\frac{\partial u}{\partial x}\right)^2 dx = -\int_0^L u\,\frac{\partial^2 u}{\partial x^2}\,dx$$

If we assume that $\partial^2 u/\partial x^2$ is bounded, the limit of this last integral must be zero as $t \to 0$. Then $\partial u/\partial t$ and $\partial u/\partial x$ are always zero, and so u is zero for all x and t.

It is not true, in general, that a solution exists. To see that this is the case, let us examine equation (W) in the region $-\infty < x < \infty$, $t > 0$, and look for a solution $u(x,t)$ which satisfies only (I). Let (x_0,t_0) be an arbitrary point, and let us consider the region in the (x,t) plane which is bounded by the following straight-line segments:

Γ_1:   passes from $(x_0 - ct_0,\ 0)$ to $(x_0 + ct_0,\ 0)$
Γ_2:   passes from $(x_0 + ct_0,\ 0)$ to (x_0,t_0)
Γ_3:   passes from (x_0,t_0) to $(x_0 - ct_0,\ 0)$

From Green's theorem,

$$\int_\Gamma\left(\frac{1}{c^2}\frac{\partial u}{\partial t}\,dx + \frac{\partial u}{\partial x}\,dt\right) = 0$$

where Γ is the closed curve consisting of Γ_1, Γ_2, and Γ_3. On Γ_1, $t = 0$ and, from (I), the integral along Γ_1 is

$$\frac{1}{c^2}\int_{x_0-ct_0}^{x_0+ct_0} g(x)\,dx$$

On Γ_2, we may replace dx by $-c\,dt$ and dt by $-1/c\,dx$. Therefore the integrand is just $-1/c\,du$, and the contribution from Γ_2 is $1/c\,[f(x_0 +$

$ct_0) - u(x_0,t_0)$]. Similarly, the contribution from Γ_3 is $1/c\ [f(x_0 - ct_0) - u(x_0,t_0)]$. Thus the solution, if it exists, is given by

$$u(x,t) = \frac{1}{2}\left[f(x + ct) + f(x - ct) + \frac{1}{c}\int_{x-ct}^{x+ct} g(x')\,dx' \right]$$

This function has first partial derivatives with respect to x and t only if f is differentiable and g is continuous. It has second partial derivatives only if f has a second derivative and if g is differentiable. Consequently, unless f and g satisfy these conditions, there is no solution to the problem.

The formal methods which we shall develop to solve the problem in which condition (B) is included will lead to certain infinite series whose values are analogous to the formula just derived. It is not surprising, therefore, that these series may not, in fact, represent a solution of the problem.

The foregoing examples are typical of the problems of mathematical physics to which the methods of this text are frequently applied. In general, we deal with a function $u(x_1, \ldots ,x_n)$ in a region R of the n-dimensional space and with certain of the partial derivatives

$$\left(\frac{\partial}{\partial x_1}\right)^{j_1}\left(\frac{\partial}{\partial x_2}\right)^{j_2} \cdots \left(\frac{\partial}{\partial x_n}\right)^{j_n} u$$

which are assumed to exist in R. If E is a function of $x_1, \ldots , x_n$ and of a set of independent variables which are identified with (and even written as) these partial derivatives, we mean by $E(u)$ the value, at each point of R, of E when its independent variables (in addition to the $x_1, \ldots , x_n$) are given the values, at $x_1, \ldots , x_n$, of u and the partial derivatives to which these independent variables correspond. If, for all points in R, $E(u) = 0$, then $u(x_1, \ldots ,x_n)$ is said to be a *solution* of the *partial-differential equation* $E(u) = 0$.

In R, $E(u) = 0$ may have a great many solutions. In the situation we generally deal with, we insist that on the boundary of R, which is an $(n - 1)$-dimensional surface, u and certain of its partial derivatives have certain specified values. If $x_1{}^2 + x_2{}^2 + \cdots + x_n{}^2$ can be made arbitrarily large, we may also impose additional conditions on u and its partial derivatives as $x_1{}^2 + x_2{}^2 + \cdots + x_n{}^2$ approaches infinity, or even as $x_1{}^2 + x_2{}^2 + \cdots x_n{}^2$ approaches infinity along certain curves or in certain sectors. The fundamental problem in the theory of partial-differential equations is to determine whether, in fact, these additional conditions specify the solution; i.e., whether or not there exists exactly one solution of the partial-differential equation which satisfies the additional conditions. Theorems which assert that there is *at most* one solution are called *uniqueness theorems;* theorems which assert that there is *at least* one solution are called *existence theorems.* If there is, for a particular situation, both uniqueness and existence, the problem is said to be *well-set*.

In each of the examples considered above uniqueness was easy to obtain. The methods discussed in the text serve to prove existence in a large number of cases; the proof is, in fact, constructive. That is, the existence of a solution is proved by exhibiting a function which can be shown to be a solution. As a bonus, whenever the solution can be constructed by these methods, uniqueness is automatically proved. To illustrate, we shall now apply these methods to the one-dimensional wave equation. (This process is illustrated in another example in Appendix C.) We shall seek a solution of (W) which satisfies (B) and (I) and for which the second partial derivatives are continuous.

From results that will be shown in Chap. 8, we can define the functions

$$u_n(x) = \sqrt{\frac{2}{L}} \sin \frac{n\pi x}{L} \qquad n = 1, 2, \ldots$$

and the symbols

$$(u,u_n) = \int_0^L u(x,t)u_n(x) \, dx$$

Then, if $\partial^2 u/\partial x^2$ is continuous,

$$\lim_{N \to \infty} \int_0^L \left[u(x,t) - \sum_{n=1}^N (u,u_n)u_n(x) \right]^2 dx = 0 \qquad (M)$$

Now, if $u(x,t)$ is a solution of (W) satisfying (B), then

$$\frac{d^2}{dt^2}(u,u_n) = \int_0^L c^2 \frac{\partial^2 u(x,t)}{\partial x^2} u_n(x) \, dx = -\frac{n^2\pi^2 c^2}{L^2}(u,u_n)$$

where the last equality is obtained by integrating by parts twice. Then

$$(u,u_n) = A \cos \frac{n\pi ct}{L} + \frac{BL}{n\pi c} \sin \frac{n\pi ct}{L}$$

where A and B are constants. Now, using (I), we obtain

$$A = (f,u_n) \qquad \text{and} \qquad B = (g,u_n)$$

This immediately proves uniqueness, for if $u(x,t)$ and $u'(x,t)$ are each solutions, then $(u,u_n) = (u',u_n)$ for all n. Using (M) in a calculation which will be verified later, we have

$$\int_0^L [u(x,t) - u'(x,t)]^2 \, dx = 0$$

which can happen only if $u(x,t) = u'(x,t)$ for all x, $0 \le x < L$, and each $t > 0$.

Now consider the partial sum

$$u_N(x,t) = \sum_{n=1}^{N} (u,u_n)u_n(x)$$

Each of the terms of this sum is a solution of (W) which satisfies (B). Consequently, $u_N(x,t)$ is a solution of (W) which satisfies (B). Furthermore,

$$\frac{\partial u_N(x,t)}{\partial t} = \sum_{n=1}^{N} \frac{d}{dt}(u,u_n)u_N(x)$$

It follows, from the results of Chap. 3, that if f and g are continuous functions, then

$$\lim_{N\to\infty} u_N(x,0) = f(x) \qquad \lim_{N\to\infty} \frac{\partial u_N(x,0)}{\partial t} = g(x)$$

Thus, if $f(x)$ and $g(x)$ are continuous, and if for each x, $0 \le x \le L$ and $t > 0$, the infinite series

$$\sum_{n=1}^{\infty} (u,u_n)u_n(x) = \lim_{N\to\infty} u_N(x,t)$$

$$\sum_{n=1}^{\infty} \frac{d}{dt}(u,u_n)u_n(x) = \lim_{N\to\infty} \frac{\partial u_N(x,t)}{\partial t}$$

converge uniformly, then the function

$$u(x,t) = \lim_{N\to\infty} u_N(x,t) = \sum_{n=1}^{\infty} (u,u_n)u_n(x)$$

satisfies (I) and (B). Although each $u_N(x,t)$ is a solution of (W), on the basis of the discussion of the one-dimensional wave equation, it should not be expected that $u(x,t)$ will be a solution of (W) unless f has a continuous second derivative and g has a continuous first derivative. But when these conditions do hold, and f and g satisfy (B), then it can be shown that $n^2(f,u_n)$ and $n(g,u_n)$ are bounded as $n \to \infty$, from which it can be demonstrated that the infinite series

$$u(x,t) = \sum_{n=1}^{\infty} (u,u_n)u_n(x)$$

has continuous second derivatives with respect to x and t and is a solution of (W).

Even when f and g do not satisfy the hypotheses required to prove the existence of a solution, the infinite series used to represent $u(x,t)$ can be given physical significance. This is achieved by using the partial

sums $u_N(x,t)$, which, as we have seen, are solutions of (W) which satisfy
(B). If f and g are just continuous, then $u_N(x,t)$ can be made to *approximately* satisfy (I), to within any error, by taking N sufficiently large.
Thus we can obtain solutions in which the functions f and g in (I) can be
replaced by the functions f_1 and g_1, where f_1 and g_1 are uniformly arbitrarily close to f and g. The use of the series

$$\sum_{n=1}^{\infty} (u,u_n)u_n(x)$$

to represent the solution is then shorthand notation for this circumstance.

The partial-differential equations for which solutions can be similarly constructed have common features. The most salient of these is
the property of *linearity*. If L is a function of $(x_1, \ldots ,x_n)$, u, and
various partial derivatives of u, and $L(u)$ is defined as was $E(u)$ above,
then $L(u)$ is said to be *linear* if for all constants α_1 and α_2, and for all
functions u_1 and u_2 having the required partial derivatives,

$$L(\alpha_1 u_1 + \alpha_2 u_2) = \alpha_1 L(u_1) + \alpha_2 L(u_2)$$

*All the partial-differential equations $E(u) = 0$ to which the methods of this
text, and generalizations of the methods of this text, can be applied are such
that*

$$E(u) = L(u) - f(x_1, \ldots ,x_n)$$

where $L(u)$ is linear.

When $f(x_1, \ldots ,x_n) \equiv 0$, we say we have a *homogeneous* linear
problem; otherwise, we say we have an *inhomogeneous* linear problem.
The partial-differential equations of mathematical physics generally
describe a relationship that must be satisfied by one or more dependent
variables $u^{(1)}, \ldots , u^{(m)}$ and various of their partial derivatives. It is
only when these relationships can be manipulated to obtain one or more
linear homogeneous or inhomogeneous problems that the solution
methods we shall discuss can be applied. Sometimes, as in the case of
the problems of electromagnetic theory, this manipulation takes place
without the introduction of additional physical hypotheses. In other
cases the manipulation is accomplished by discarding those parts of the
partial-differential equation which cause it not to be linear. This process
is called a *linearization* of the problem. In some instances the linearization is disguised. For example, the equations of elasticity appear to be
linear, but this is because of an earlier assumption of the linearity of the
relationship between stress and strain, or because of the approximation
of certain functions by the linear and quadratic terms of their power-
series expansion. The effect of the linearity is felt in several ways,
depending upon whether the problem is homogeneous or inhomogeneous

and the nature of the additional conditions that the solution is required to satisfy.

For the inhomogeneous problem $L(u) = f$, suppose that all the additional conditions are *linear;* that is, if u_1 and u_2 satisfy them, so does $\alpha_1 u_1 + \alpha_2 u_2$. Condition (B) for the case of the one-dimensional wave equation represents linear conditions. Then the method of solution is to attempt to find a sequence of functions $u_1, u_2, \ldots, u_n, \ldots$ having the property that

$$L(u_n) = \lambda_n u_n$$

where λ_n is a constant, and to seek a solution of the form

$$u = \sum_n c_n u_n$$

where the c_n are constants. If we neglect questions of convergence, it is clear from the linearity that if each of the u_n satisfies the additional conditions, then so does u, and that

$$L(u) = \sum_n c_n \lambda_n u_n$$

Then, if f can be represented as a series

$$f = \sum_n f_n u_n$$

f_n constant, and if none of the λ_n is zero, a solution can be found by taking $c_n = f_n/\lambda_n$.

For the homogeneous problem $L(u) = 0$, the situation is different. Here we have some conditions which are linear and some which are not. To distinguish between them the linear conditions are called *boundary conditions* and the others are called *initial conditions.* Condition (I) for the one-dimensional wave equation represents initial conditions. This terminology is especially appropriate when one of the independent variables is time t, the other independent variables describe a region in space, the initial conditions describe the behavior of the function and its time derivatives, where $t = 0$, and the boundary conditions describe conditions that must be satisfied by the function and its spatial derivatives at the boundary of the spatial region. The method of attack here is to find a sequence of solutions u_n for which $L(u_n) = 0$ and which satisfy the boundary conditions. Again, we seek a solution of the form $u = \sum_n c_n u_n$, where the c_n are constants. Again, neglecting questions of convergence, the linearity assures that $L(u) = 0$ and u satisfies the boundary conditions. The problem can then be solved if the c_n can be found such that u satisfies the initial conditions.

The correct mathematical setting for a deeper discussion of these problems is based on the theories of topological groups and topological vector spaces.   This book stops short of such a venture.

SOME OTHER TOPICS

The physical applications chosen for discussion are those which, for the most part, lead to problems that can be solved within the framework of the text material.   There are, however, several important subjects that extend slightly beyond our charted territory.

The nonrelativistic quantum theory of a single particle is such a subject.   For such a particle, the appropriate partial-differential equation is

$$-\frac{h^2}{8m\pi^2}\nabla^2\Psi + V(x,y,z)\Psi = \frac{ih}{2\pi}\frac{\partial\Psi}{\partial t} \tag{S}$$

This is called *Schrödinger's equation*.   The constants m and h are the mass of the particle and Planck's constant.   The complex-valued function $\Psi(x,y,z,t)$ is called the *wave function*.   Its physical interpretation is that $|\Psi|^2 = \Psi\,\tilde{\Psi}$ ($\tilde{\Psi}$ is the complex conjugate of Ψ) is the probability density for the particle.   That is, in any region R of the three-dimensional space, the probability at time t that the particle lies in R is

$$\int_R |\Psi|^2\, dR$$

The function $V(x,y,z)$ is the classical potential energy of the particle.

The left-hand member of (S) can be thought of as $L(\Psi)$, where L is linear.   Then

$$L(\Psi) = \lambda\Psi$$

is analogous to the Sturm-Liouville equation, which we shall study in Chap. 8.   When Ψ and V depend only upon x, it is precisely the Sturm-Liouville equation.   However, replacing the homogeneous boundary conditions or the periodic conditions is the condition that $|\Psi|^2$ be integrable over the entire three-dimensional space.   The interpretation of $|\Psi|^2$ as a probability distribution, in fact, requires that

$$\int\!\!\int\!\!\int_{-\infty}^{\infty} |\Psi(x,y,z,t)|^2\, dx\, dy\, dz = 1$$

Now, for certain functions $V(x,y,z)$, the basic results of the Sturm-Liouville theory continue to apply.   There exists a sequence of functions

Ψ_n and real constants λ_n such that

$$L(\Psi_n) = \lambda_n \Psi_n$$

$$(\Psi_n, \Psi_m) = \begin{cases} 1 & \text{if } n = m \\ 0 & \text{if } n \neq m \end{cases}$$

where

$$(\Psi_n, \Psi_m) = \iiint_{-\infty}^{\infty} \Psi_n \tilde{\Psi}_m \, dx \, dy \, dz$$

and, if $|L(\Psi)|^2$ is integrable over all space,

$$\lim_{N \to \infty} \iiint_{-\infty}^{\infty} \left| \Psi - \sum_{n=1}^{N} (\Psi, \Psi_n)\Psi_n \right|^2 dx \, dy \, dz = 0 \tag{M}$$

Then, using the same technique as is used for the heat equation (Appendix C), we find that

$$(\Psi, \Psi_n) = a_n e^{-2\pi i \lambda_n t / h}$$

where a_n is a constant. Each of the Ψ_n is called a *state* of the particle, and the number $|a_n|^2$ is interpreted as the probability that the particle is in its nth state when $t = 0$.

In the more general situation, not treated in this text, either there do not exist any λ_n, or if there do, the equality (M) is not satisfied. As an example of this, suppose that $V = 0$ and that the particle is known to lie on the x axis. Then (S) becomes

$$\frac{\partial^2 \Psi}{\partial x^2} = -\frac{4\pi i m}{h} \frac{\partial \Psi}{\partial t} \tag{S'}$$

and

$$\int_{-\infty}^{\infty} |\Psi|^2 \, dx = 1$$

Then, using the theory of the Fourier integral (Chap. 6) and ignoring questions of convergence, we replace (M) by the Fourier integral

$$\Psi(x,t) = \frac{1}{2\pi} \int_{-\infty}^{\infty} \hat{\Psi}(\omega,t) e^{i\omega x} \, d\omega$$

Then, using (S'), we have

$$\frac{d\hat{\Psi}}{dt} = -\frac{ih\omega^2}{4\pi m} \hat{\Psi}$$

or

$$\hat{\Psi} = a(\omega) e^{-ih\omega^2 t / 4\pi m}$$

In the case in which (M) applies, the system is said to have a *discrete spectrum*. In the case above, that of a free particle, the system is said to have a *continuous spectrum*. In the most general cases, the system may have both a continuous and a discrete spectrum; a representation for $\Psi(x,t)$ analogous to the Fourier integral can be formed.

In the theories of probability and statistics, the Fourier transform becomes a powerful tool because of the convolution property for the sum of random variables. This property is that if f_1 and f_2 are independent random variables having distribution functions $D_1(x)$ and $D_2(x)$, then the distribution function for $f_1 + f_2$ is

$$D(x) = \int_{-\infty}^{\infty} D_1(x - y) D_2(y) \, dy$$

Using the results of Chap. 6, we can rewrite this function in terms of Fourier transforms as

$$\hat{D}(\omega) = \hat{D}_1(\omega) \hat{D}_2(\omega)$$

This fact is useful in the analysis of a large number of independent random variables.

1
Periodic Functions

DEFINITION OF PERIODIC FUNCTIONS

Let f be a function defined for all real values of t, and let $L > 0$ be such that for all t,

$$f(t + L) = f(t)$$

The number L is then said to be a *period* of f, and f is said to be *periodic*, of period L. If L is a period of f, it is clear that

$$f(t + nL) = f(t)$$

for all integers n and for all t. It is also clear that for any r, L is a period for the function $g(t) = f(t + r)$, and that if f and g are both periodic, of period L, then so are their product fg and their sum $f + g$.

Theorem 1 Let f be periodic, of period L. Suppose that for some t_0 the integral

$$\int_{t_0}^{t_0+L} f(t)\, dt$$

exists.   Then for every real value of r the integral

$$\int_{t_0}^{t_0+L} f(t+r)\, dt$$

exists, and for every t_1 the integral

$$\int_{t_1}^{t_1+L} f(t)\, dt$$

exists, and all these integrals have the same value.

Proof   Since there is a unique integer n and a unique r_1, $0 \le r_1 < L$, such that $r = nL + r_1$, we lose no generality by assuming that $0 < r < L$.   Now,

$$\int_{t_0}^{t_0+L} f(t)\, dt = \int_{t_0}^{t_0+r} f(t)\, dt + \int_{t_0+r}^{t_0+L} f(t)\, dt$$
$$= \int_{t_0}^{t_0+r} f(t+L)\, dt + \int_{t_0+r}^{t_0+L} f(t)\, dt$$

If we make the change of variables $\tau = t + L - r$ and $\tau = t - r$, respectively, in the last two integrals, they become

$$\int_{t_0+L-r}^{t_0+L} f(\tau+r)\, dt \qquad \text{and} \qquad \int_{t_0}^{t_0+L-r} f(\tau+r)\, dt$$

so that their sum is

$$\int_{t_0}^{t_0+L} f(t+r)\, dt$$

Using this result, we have

$$\int_{t_0}^{t_0+L} f(t)\, dt = \int_{t_0}^{t_0+L} f(t+(t_1-t_0))\, dt = \int_{t_1}^{t_1+L} f(t)\, dt$$

EXTENSIONS

Let us suppose that f is periodic, of period L.   We observe that if the values of f are known for all values of t lying on an interval $t_0 \le t < t_0 + L$, then the values of f for all other values of t can be determined by the following device.   For any t_1 there is a unique integer n and a unique r, $0 \le r < L$, such that $t_1 - t_0 = nL + r$.   Therefore

$$f(t_1) = f(t_0 + r + nL) = f(t_0 + r)$$

Conversely, suppose that a function f is defined just for the values of t for which $t_0 \le t < t_0 + L$.   Suppose that for arbitrary t_1 we define $\tilde{f}(t_1)$ by the above formula.   Then $\tilde{f}(t_1)$ is defined for all t_1 as $\tilde{f}(t_1 + L) = \tilde{f}(t_1)$ and for $t_0 \le t_1 < t_0 + L$ as $\tilde{f}(t_1) = f(t_1)$. $\tilde{f}$ is then said to be the *periodic extension* of f.   We shall often deal with functions defined on an interval $t_0 \le t < t_0 + L$, and we shall often study such functions by obtaining properties of their periodic extensions.

FOURIER SERIES

Consider the sequence of functions

$$\phi_n(t) = \frac{1}{\sqrt{L}} e^{2\pi int/L} = \frac{1}{\sqrt{L}}\left(\cos\frac{2\pi nt}{L} + i\sin\frac{2\pi nt}{L}\right)$$

$$n = 0, \pm 1, +2, \ldots$$

These functions are periodic, of period L, and play a central role in the theory of periodic functions. In order to study this role in an efficient fashion, we introduce the notation

$$(f,g) = \int_0^L f(t)\bar{g}(t)\,dt$$

in which f and g are integrable functions on the interval $[0,L]$ and the bar indicates complex conjugation. It is then easy to verify that

$$(\phi_n,\phi_m) = \delta_{nm}$$

where

$$\delta_{nm} = \begin{cases} 1 & \text{if } n = m \\ 0 & \text{if } n \neq m \end{cases}$$

Let f be periodic, of period L, and let f be integrable on $[0,L]$. Then the numbers

$$(f,\phi_n) \qquad n = 0, \pm 1, \pm 2, \ldots$$

are called the *Fourier coefficients* of f, and the infinite series

$$\sum_{n=-\infty}^{\infty} (f,\phi_n)\phi_n$$

is called the *Fourier series* for f. In general, *the Fourier series may not converge* and must be regarded as a purely formal expression. To emphasize this state of affairs, we write

$$f \sim \sum_{n=-\infty}^{\infty} (f,\phi_n)\phi_n$$

where the symbol $\sim$ expresses the fact that the Fourier series on the right is determined by f but is not necessarily a convergent series; nor, in case it does converge, does it necessarily converge to f. It is sometimes convenient to replace the Fourier series by another series, using the real trigonometric functions instead of the complex exponential functions.

This is accomplished in the following manner:

$$(f, \phi_n)\phi_n = \frac{1}{L} \int_0^L f(\tau) \left(\cos \frac{2n\pi\tau}{L} - i \sin \frac{2\pi n\tau}{L} \right) d\tau$$

$$\times \left(\cos \frac{2\pi nit}{L} + i \sin \frac{2\pi nit}{L} \right)$$

Then

$$(f, \phi_0)\phi_0 = \frac{1}{L} \int_0^L f(\tau) \, d\tau$$

and

$$(f, \phi_n)\phi_n + (f, \phi_{-n})\phi_{-n} = \left[\frac{2}{L} \int_0^L f(\tau) \cos \frac{2\pi n\tau}{L} \, d\tau \right] \cos \frac{2\pi nt}{L}$$

$$+ \left[\frac{2}{L} \int_0^L f(\tau) \sin \frac{2\pi n\tau}{L} \, d\tau \right] \sin \frac{2\pi nt}{L}$$

Thus the Fourier series for f can be written in the form

$$\frac{a_0}{2} + \sum_{n=1}^{\infty} \left(a_n \cos \frac{2\pi nt}{L} + b_n \sin \frac{2\pi nt}{L} \right)$$

where

$$a_n = \frac{2}{L} \int_0^L f(t) \cos \frac{2\pi nt}{L} \, dt \qquad n = 0, 1, 2, \ldots$$

and

$$b_n = \frac{2}{L} \int_0^L f(t) \sin \frac{2\pi nt}{L} \, dt \qquad n = 0, 1, 2, \ldots$$

When written in this form, the Fourier series is said to be in *real trigonometric form*.

Problem   Find the Fourier coefficients for the following functions, and find the real trigonometric form of their Fourier series:

(a) $f(t) = \sin \dfrac{t}{2} \qquad 0 \leq t \leq 2\pi$

(b) $f(t) = \dfrac{t(2\pi - t)}{\pi^2} \qquad 0 \leq t \leq 2\pi$

(c) $f(t) = \pi - t \qquad 0 \leq t < 2\pi$

(d) $f(t) = \begin{cases} \pi & 0 \leq t < \pi \\ -\pi & \pi \leq t < 2\pi \end{cases}$

Answers

(a) $(f,\phi_n) = \dfrac{4}{\sqrt{2\pi}} \dfrac{1}{1 - 4n^2}$

$f \sim \dfrac{2}{\pi} - \dfrac{4}{\pi} \displaystyle\sum_{n=1}^{\infty} \dfrac{\cos nt}{4n^2 - 1}$

(b) $(f,\phi_0) = \dfrac{4}{3}\sqrt{\dfrac{\pi}{2}}$ $\quad (f,\phi_n) = -\left(\dfrac{2}{\pi}\right)^{3/2} \dfrac{1}{n^2} \quad n \neq 0$

$f \sim \dfrac{2}{3} - \dfrac{4}{\pi^2} \displaystyle\sum_{n=1}^{\infty} \dfrac{\cos nt}{n^2}$

(c) $(f,\phi_0) = 0$ $\quad (f,\phi_n) = \dfrac{\sqrt{2\pi}}{in} \quad n \neq 0$

$f \sim 2 \displaystyle\sum_{n=1}^{\infty} \dfrac{\sin nt}{n}$

(d) $(f,\phi_n) = \begin{cases} 4\sqrt{\dfrac{\pi}{2}}\dfrac{1}{in} & n \text{ odd} \\ 0 & n \text{ even} \end{cases}$

$f \sim 2 \displaystyle\sum_{n=0}^{\infty} \dfrac{\sin (2n + 1)t}{2n + 1}$

BESSEL'S INEQUALITY AND PARSEVAL'S EQUALITY

We shall now prove two theorems which provide a little insight into the nature of Fourier series.

Theorem 2 Let

$$\sum_{n=-\infty}^{\infty} c_n \phi_n$$

converge uniformly on $[0,L]$; that is, let the sequence of partial sums

$$S_N(t) = \sum_{n=-N}^{N} c_n \phi_n(t)$$

converge uniformly to a function $f(t)$. Then $c_n = (f,\phi_n)$.

Proof For a uniformly convergent sequence of functions, the integral of the limit is the limit of the integrals. Since

$$S_N(t)\,\bar{\phi}_m(t) = \sum_{m=-N}^{N} c_n \phi_n(t)\,\bar{\phi}_m(t)$$

converges uniformly to $f(t)\,\bar{\phi}_m(t)$, then

$$(f,\phi_m) = \lim_{N\to\infty} \sum_{n=-\infty}^{\infty} c_m \delta_{nm} = c_m$$

This theorem does *not* imply that if the Fourier series for f converges uniformly, it must converge to f.   All that it says is that if the Fourier series for f converges uniformly to a function, say g, then $(g,\phi_n) = (f,\phi_n)$ for all n.   Since (f,ϕ_n) is not altered by changing the value of $f(t)$ at a finite number of values of t, it is clear that this set of equalities does not imply that f and g are the same function.   This question will be investigated in the next two chapters, which culminate in a definitive theorem.

Theorem 3   Consider all functions

$$F_N = \sum_{n=-N}^{N} c_n \phi_n$$

where c_n are complex numbers.   Among functions of this type, the one which minimizes the integral

$$(F_N - f, F_N - f) = \int_0^L |F_n(t) - f(t)|^2\, dt$$

is the one for which

$$c_n = (f,\phi_n) \qquad n = 0, \pm 1, \ldots, \pm N$$

Proof   We first observe that

$$(f + g, h) = (f,h) + (g,h) \qquad (f,g) = (\bar{g},\bar{f}) \qquad (cf,g) = c(f,g)$$

for integrable f, g, and h and constant c.   Then

$$(F_N - f, F_N - f) = (F_N,F_N) - (F_n,f) - (f,F_N) + (f,f)$$

$$= \left(\sum_{n=-N}^{N} C_n \phi_n, \sum_{n=-N}^{N} C_n \phi_n \right) - \sum_{n=-N}^{N} C_n (\phi_n,f)$$

$$- \sum_{n=-N}^{N} \bar{C}_n (f,\phi_n) + (f,f)$$

$$= \sum_{n=-N}^{N} [C_n\bar{C}_n(\phi_n,f) - \bar{C}_n(f,\phi_n)] + (f,f)$$

$$= \sum_{n=-N}^{N} |C_n - (f,\phi_n)|^2 + (f,f) - \sum_{n=-N}^{N} |(f,\phi_n)|^2$$

Since $(F_N - f, F_N - f) \geq 0$, this expression assumes its minimum value when $C_n = (f,\phi_n)$, $n = 0, \pm 1, \ldots, \pm N$.

If the C_n in the expression for F_N in this theorem are given the values (f,ϕ_n), then we obtain the equality

$$\sum_{n=-N}^{N} |(f,\phi_n)|^2 = (f,f) - \Big[\sum_{n=-N}^{N} (f,\phi_n)\phi_n - f, \sum_{n=-N}^{N} (f,\phi_n)\phi_n - f\Big]$$

Since

$$\sum_{n=-N}^{N} |(f,\phi_n)|^2 \leq (f,f) \qquad \text{for all } N$$

the limit as $N \to \infty$ exists, and we have *Bessel's inequality*,

$$\sum_{n=-\infty}^{\infty} |(f,\phi_n)|^2 \leq (f,f)$$

Furthermore,

$$\lim_{N\to\infty} \int_0^L \Big|\Big[\sum_{n=-N}^{N} (f,\phi_n)\phi_n - f\Big]\Big|^2 dt = 0$$

if and only if *Parseval's equality*,

$$\sum_{n=-\infty}^{\infty} |(f,\phi_n)|^2 = (f,f)$$

is satisfied. In Chap. 4 we shall see the circumstances under which Parseval's equality is satisfied and what its consequences are.

Problem

(a) If

$$(f,g) = \int_{-1}^{1} f(t)g(t)\, dt$$

and if

$$\psi_1 = \frac{1}{\sqrt{2}} \qquad \psi_2 = \sqrt{\frac{3}{2}} t \qquad \psi_3 = \frac{1}{2}\sqrt{\frac{5}{2}}(3t^2 - 1)$$

show that

$(\psi_1,\psi_2) = (\psi_1,\psi_3) = (\psi_2,\psi_3) = 0$
$(\psi_1,\psi_1) = (\psi_2,\psi_2) = (\psi_3,\psi_3) = 1$

(b) Find the second-degree polynomial $g(t)$ such that

$$\int_{-1}^{1} \Big[\cos\frac{\pi t}{2} - g(t)\Big]^2 dt$$

is as small as possible.

Answer

$$g(t) = \frac{3}{\pi^3}[(20 - \pi^2) - 5(12 - \pi^2)t^2]$$

2
The Cesaro Sum

SEQUENCE OF AVERAGES

We have obtained some preliminary connections between an integrable periodic function and its Fourier series. Let us now digress to some further connections. If $A_0, A_1, \ldots, A_n, \ldots$ is a sequence of numbers, it may be that the sequence diverges, but that the sequence of averages

$$T_1 = A_0, \quad T_2 = \frac{A_0 + A_1}{2}, \ldots$$

$$T_n = \frac{A_0 + A_1 + \cdots + A_{n-1}}{n}, \ldots$$

nevertheless converges. For example, if $A_n = (-1)^n$, the sequence diverges, but the sequence of averages, $1, 0, \frac{1}{3}, 0, \frac{1}{5}, \ldots$, converges to zero. However, if the sequence has a limit A, the sequence of averages must converge to A. We state this as a theorem.

Theorem 1   Let A_0, A_1, . . . , A_n, . . . be a convergent sequence,

$$\lim_{n \to \infty} A_n = A$$

If

$$T_n = \frac{A_0 + A_1 + \cdots + A_{n-1}}{n}$$

then

$$\lim_{n \to \infty} T_n = A$$

Proof   Let $\epsilon > 0$.   Then let $N(\epsilon)$ be such that if $n > N(\epsilon)$,

$$|A_n - A| < \frac{\epsilon}{2}$$

If $p > 0$,

$$T_{N+p} - A = \frac{A_0 + A_1 + \cdots + A_{N-1}}{N + p}$$
$$+ \frac{A_N + A_{N+1} + \cdots + A_{N+p-1}}{N + p} - A$$

$$= \frac{A_0 + \cdots + A_{N-1}}{N + p}$$
$$+ \frac{(A_N - A) + \cdots + (A_{N+p-1} - A)}{N + p} - \frac{NA}{N + p}$$

If $N = N(\epsilon)$,

$$|T_{N+p} - A| < \left| \frac{A_0 + \cdots + A_{N-1}}{N + p} \right| + \frac{N}{N + p}|A| + \frac{p}{N + p}\frac{\epsilon}{2}$$

If p is sufficiently large,

$$|T_{N+p} - A| < \epsilon$$

THE CESARO SUM

This theorem can be translated into the language of series.

Theorem 2   Let

$$\sum_{n=-\infty}^{\infty} a_n$$

be a series which converges to a value A.   Then

$$\lim_{n \to \infty} \sum_{j=-(n-1)}^{n-1} \left(1 - \frac{|j|}{n} \right) a_j = A$$

Proof If we set

$$A_0 = a_0, \; A_1 = a_1 + a_0 + a_1, \; \ldots, \; A_n = a_{-n} + \cdots + a_n$$

the hypothesis of the preceding theorem is satisfied, and

$$T_n = \sum_{j=-(n-1)}^{(n-1)} \left(1 - \frac{|j|}{n}\right) a_j$$

converges to A.

If the series

$$\sum_{n=-\infty}^{\infty} a_n$$

does not converge, and if instead

$$\lim_{n \to \infty} \sum_{j=-(n+1)}^{(n+1)} \left(1 - \frac{|j|}{n}\right) a_j = T$$

then we call T the *first Cesaro sum* of the series. We see that when a series converges, its first Cesaro sum exists and is equal to the limit of the series; however, the first Cesaro sum may exist without the series converging.

THE ABEL SUM

If

$$b_0 + b_1 + \cdots + b_n + \cdots$$

is a divergent series, but the numbers b_n remain bounded, then the series

$$b_0 + b_1, \; x + \cdots + b_n x^n + \cdots$$

converges if $0 < x < 1$. Let this series converge to $B(x)$. It is possible that as $x \to 1$ through real values less than 1, $B(x)$ has a limiting value B. This number,

$$B = \lim_{x \to 1^-} B(x)$$

is called the *Abel sum* of the original series; for example, the divergent series

$$\sum_{n=0}^{\infty} (-1)^n$$

produces

$$B(x) = \frac{1}{1+x}$$

so that its Abel sum is $\frac{1}{2}$.   If it is true that the Abel sum of a convergent series is the same as the value of the series, we have another method of assigning a number to a divergent series which is consistent, in the case of a convergent series, with ordinary convergence.   The next theorem shows that this is, in fact, true.

Theorem 3   Let the series

$$\sum_{n=0}^{\infty} b_n$$

converge.   Then

$$\lim_{x \to 1^-} \sum_{n=0}^{\infty} b_n x^n = \sum_{n=0}^{\infty} b_n$$

Proof   Let $\epsilon > 0$.   Then there is a J such that if $j > J$,

$$\left| \sum_{n=j+1}^{j+p} b_n \right| < \frac{\epsilon}{2} \qquad \text{for all } p$$

Let $N > J$.   Then

$$\sum_{n=0}^{N} b_n - \sum_{n=0}^{N} b_n x^n = \sum_{n=1}^{J} (1 - x^n) b_n + (1 - x) \sum_{n=J+1}^{N} b_n \sum_{j=0}^{n-1} x^j$$

$$= \sum_{n=1}^{J} (1 - x^n) b_n + (1 - x) \left(\sum_{j=0}^{J} x^j \sum_{n=J+1}^{N} b_n \right.$$

$$\left. + \sum_{j=J}^{N-1} x^j \sum_{n=j+1}^{N} b_n \right)$$

Thus

$$\left| \sum_{n=0}^{N} b_n - \sum_{n=0}^{N} b_n x^n \right| < \left| \sum_{n=1}^{J} (1 - x^n) b_n \right| + (1 - x^N) \frac{\epsilon}{2}$$

This inequality is valid for all $N > J$, and therefore for all x, $0 < x < 1$,

$$\left| \sum_{n=0}^{\infty} b_n - \sum_{n=0}^{\infty} b_n x^n \right| < \left| \sum_{n=1}^{J} (1 - x^n) b_n \right| + \frac{\epsilon}{2}$$

Now, if x is sufficiently close to 1, the right-hand member of this inequality is less than ϵ.

For a series

$$\sum_{n=-\infty}^{\infty} a_n$$

the Abel sum is defined as the Abel sum of

$$\sum_{n=0}^{\infty} b_n$$

where

$$b_0 = a_0 \quad \text{and} \quad b_n = a_n + a_{-n} \quad n \geq 1$$

Problem
   (a) Show that the infinite series

$$\sum_{n=-\infty}^{\infty} e^{int}$$

does not converge for any value of t, but that if $0 < t < 2\pi$, it has a first Cesaro sum equal to zero.
   (b) Show that if $0 < t < 2\pi$, this series has an Abel sum equal to zero.

Problem   Show that the infinite series

$$\sum_{n=1}^{\infty} (-1)^n n$$

does not converge, does not have a first Cesaro sum, and has an Abel sum equal to $-\frac{1}{4}$.

3
The General Question

KERNEL SEQUENCES

If an integrable function f, with period L, is given, its Fourier coefficients (f, ϕ_n) are determined. The question we are now concerned with is this: If the Fourier coefficients are given, how can f be determined? We have already observed that the values of f can be changed at a finite number of points without changing the values of (f, ϕ_n). Therefore it is trivially obvious that different functions can have the same Fourier coefficients. However, let us assume that at some point t_0, f has both a right-hand and left-hand limit, designated, respectively, as $f(t_{0+})$ and $f(t_{0-})$. What we shall show is that the number

$$\tfrac{1}{2}[f(t_{0+}) + f(t_{0-})]$$

is completely determined by the Fourier coefficients and may be obtained by computing either the Cesaro sum or the Abel sum of the Fourier series

$$\sum_{n=-\infty}^{\infty} (f, \phi_n) \phi_n(t_0)$$

In particular, if f is continuous at t_0, then

$$f(t_{0+}) = f(t_{0-}) = f(t_0)$$

so that

$$\tfrac{1}{2}[f(t_{0+}) + f(t_{0-})] = f(t_0)$$

If f is continuous for all values of t, it is then completely determined by its Fourier coefficients.

Let $\{K_n\}$ be a sequence of functions of the real variable t, having the following properties:

Each K_n is a continuous function of t, with period L, and
$$K_n(-t) = K_n(t)$$ (K-1)

$$K_n(t) \geq 0 \text{ for all } t$$ (K-2)

$$\int_0^L K_n(t)\, dt = 1 \text{ for all } n$$ (K-3)

If $0 < t_1 < L/2$, for every $\epsilon > 0$ there exists an N such that if $n > N$, $|(K_n(t)| < \epsilon$ for all t in the interval $t_1 \leq t \leq L/2$ (K-4)

Such a sequence of functions will be called a *kernel sequence*.

Theorem 1   Let $\{K_n\}$ be a kernel sequence.   Let f be an integrable function, with period L.   If t_0 is a point at which f has both right-hand and left-hand limits, then

$$\lim_{n \to \infty} \int_0^L K_n(t - t_0)f(t)\, dt = \tfrac{1}{2}[f(t_{0+}) + f(t_{0-})]$$

If f is continuous in an interval $t_1 \leq t \leq t_2$, then not only does

$$\lim_{n \to \infty} \int_0^L K_n(t - t_0)f(t)\, dt = f(t_0) \qquad \text{for all } t_0, t_1 \leq t_0 \leq t_2$$

but the convergence is uniform on that interval.

Proof   The integral

$$\int_0^L K_n(t - t_0)f(t)\, dt$$

can be rewritten as

$$\int_{-t_0}^{L-t_0} K_n(\tau)f(t_0 + \tau)\, d\tau$$

and this integral, in turn, is the same as

$$\int_{-L/2}^{L/2} K_n(\tau)f(t_0 + \tau)\, d\tau = \int_{-L/2}^{0} K_n(\tau)f(t_0 + \tau)\, d\tau$$
$$+ \int_0^{L/2} K_n(\tau)f(t_0 + \tau)\, d\tau$$

Let us consider the integral

$$\int_0^{L/2} K_n(\tau) f(t_0 + \tau) \, d\tau = f(t_{0+}) \int_0^{L/2} K_n(\tau) \, d\tau$$
$$+ \int_0^{L/2} K_n(\tau) [f(t_0 + \tau) - f(t_{0+})] \, d\tau$$

Because

$$\int_{-L/2}^{L/2} K_n(t) \, dt = \int_0^L K_n(t) \, dt = 1 \qquad \text{and} \qquad K_n(t) = K_n(-t)$$

we obtain

$$\int_0^{L/2} K_n(\tau) \, d\tau = \tfrac{1}{2}$$

We have now to prove that

$$\lim_{n \to \infty} \int_0^{L/2} K_n(\tau) [f(t_0 + \tau) - f(t_{0+})] \, d\tau = 0$$

Let $\epsilon > 0$.   There exists a t_1, $0 < t_1 < L/2$, such that for $0 < \tau < t_1$,

$$|f(t_0 + \tau) - f(t_{0+})| < \epsilon$$

If f is continuous in a closed interval, this t_1 can be chosen to be the same for every t_0 in the interval.   Then

$$\int_0^{L/2} K_n(\tau) [f(t_0 + \tau) - f(t_{0+})] \, d\tau = \int_0^{t_1} K_n(\tau) [f(t_0 + \tau) - f(t_{0+})] \, d\tau$$
$$+ \int_{t_1}^{L/2} K_n(\tau) [f(t_0 + \tau) - f(t_{0+})] \, d\tau$$

But

$$\left| \int_0^{t_1} K_n(\tau) [f(t_0 + \tau) - f(t_{0+})] \, d\tau \right| < \int_0^{t_1} |K_n(\tau)| \epsilon \, d\tau$$
$$= \epsilon \int_0^{t_1} K_n(\tau) \, d\tau \leq \epsilon \int_0^{L/2} K_n(\tau) \, d\tau = \frac{\epsilon}{2}$$

Let M be an upper bound for $|f|$ (since f is integrable, it is bounded), and let N be such that for $n > N$ and $t_1 \leq t \leq L/2$,

$$K_n(t) < \frac{1}{LM} \frac{\epsilon}{2}$$

Then

$$\left| \int_{t_1}^{L/2} K_n(\tau) [f(t_0 + \tau) - f(t_{0+})] \, d\tau \right| < \left(\frac{L}{2} - t_1 \right) \frac{1}{L} \epsilon < \frac{\epsilon}{2}$$

We have then exhibited an N such that if $n > N$,

$$\left| \int_0^{L/2} K_n(\tau) [f(t_0 + \tau) - f(t_{0+})] \, d\tau \right| < \epsilon$$

Furthermore, if f is continuous on a closed interval, this N is the same for every t_0 on the interval.   A similar argument can be applied to the integral

$$\int_{-L/2}^{0} K_n(\tau) f(t_0 + \tau)\, d\tau$$

where $f(t_{0^+})$ is replaced by $f(t_{0^-})$.

THE FEJER KERNELS

We shall now find that computing the Cesaro sum of a Fourier series is equivalent to computing

$$\lim_{n \to \infty} \int_0^L K_n(t - t_0) f(t)\, dt$$

where K_n is a kernel sequence.

The nth partial sum of the Fourier series for f, evaluated at t_0, is

$$
\begin{aligned}
A_n(t_0) &= \sum_{k=-n}^{n} (f, \phi_k) \phi_k(t_0) \\
&= \sum_{k=-n}^{n} \frac{1}{L} \left[\int_0^L f(t) e^{-(2\pi k i/L) t}\, dt \right] e^{(2\pi i k/L) t_0} \\
&= \int_0^L \frac{1}{L} \left(\sum_{k=-n}^{n} e^{(2\pi k i/L)(t_0 - t)} \right) f(t)\, dt
\end{aligned}
$$

Let

$$D_n(t) = \frac{1}{L} \sum_{k=-n}^{n} e^{(2\pi k i/L) t}$$

It is clear that

$$\int_0^L D_n(t)\, dt = 1$$

For $t = 0$,

$$D_n(0) = \frac{2n + 1}{L}$$

For $0 < t < L$, let $\zeta = 2\pi t/L$. Then

$$
\begin{aligned}
D_n(t) &= \frac{1}{L} \left[1 + \sum_{k=1}^{n} (e^{i\zeta})^k + \sum_{k=1}^{n} (e^{-i\zeta})^k \right] \\
&= \frac{1}{L} \left[1 + \frac{e^{i\zeta}(1 - e^{in\zeta})}{1 - e^{i\zeta}} + \frac{e^{-i\zeta}(1 - e^{-in\zeta})}{1 - e^{-i\zeta}} \right] \\
&= \frac{1}{L} \left[1 - \frac{e^{i\zeta/2}(1 - e^{in\zeta})}{e^{i\zeta/2} - e^{-i\zeta/2}} + \frac{e^{i\zeta/2}(1 - e^{-in\zeta})}{e^{i\zeta/2} - e^{-i\zeta/2}} \right] \\
&= \frac{1}{L} \frac{\sin\left(n + \frac{1}{2}\right)\zeta}{\sin(\zeta/2)} = \frac{1}{L} \frac{\sin\,(2n + 1)\pi t/L}{\sin\,(\pi t/L)}
\end{aligned}
$$

Since $D_n(t)$ is obviously periodic, of period L, and this last expression both is periodic, of period L, and has $D_n(0)$ on its limit when $t \to 0$,

$$D_n(t) = \frac{1}{L} \frac{\sin (2n+1)\pi t/L}{\sin (\pi t/L)} \qquad \text{for all } t$$

and

$$A_n(t_0) = \int_0^L D_n(t - t_0) f(t)\, dt$$

Now, the Cesaro sum is found by forming the averages

$$T_n = \frac{1}{n}(A_0 + A_1 + \cdots + A_{n-1})$$

Thus

$$T_n(t_0) = \int_0^L \left[\frac{1}{n} \sum_{j=0}^{n-1} D_j(t - t_0) \right] f(t)\, dt$$

Let

$$K_n(t) = \frac{1}{n} \sum_{j=0}^{n-1} D_j(t)$$

We shall show that $\{K_n\}$ forms a Kernel sequence. Property (K-1) is satisfied by each of the D_j, and hence by K_n. Because

$$\int_0^L D_j(t)\, dt = 1$$

property (K-3) is satisfied by K_n. In order to show that properties (K-2) and (K-4) are satisfied, another representation for $K_n(t)$ must be found. Let us write

$$K_n(t) = \frac{1}{nL \sin (\pi t/L)} \sum_{j=0}^{n-1} \sin (2j+1) \frac{\pi t}{L}$$

$$= \frac{1}{nL \sin (\pi t/L)} \sum_{j=0}^{n-1} \frac{1}{2i} [e^{\pi it/L}(e^{2\pi it/L})^j - e^{-\pi it/L}(e^{-2\pi it/L})^j]$$

For $t = 0$,

$$K_n(0) = \frac{1}{n} \sum_{j=0}^{n-1} \frac{2j+1}{L} = \frac{n}{L}$$

For $0 < t \leq L/2$ or $-L/2 \leq t < 0$ the geometric progressions can be simplified, as before; thus we obtain

$$K_n(t) = \frac{1}{nL} \left[\frac{\sin (\pi n t/L)}{\sin (\pi t/L)} \right]^2$$

If this last expression is identified at $t = 0$ with its limit as $t \to 0$, it then coincides with $K_n(t)$ on the interval $-L/2 \le t \le L/2$, and since both it and $K_n(t)$ are periodic, with period L, it must coincide with $K_n(t)$ for all t. It is now obvious that (K-3) is satisfied. For $0 < t_1 < L/2$ and $t_1 \le t \le L/2$,

$$|K_n(t)| \le \frac{1}{nL} \frac{1}{\sin^2 (\pi t_1/L)}$$

and so the K_n satisfy (K-4).

We may now state the *Fejer theorem* as follows.

Theorem 2 Let f be an integrable function, of period L. Let t_0 be a point at which f has both a right-hand and a left-hand limit. Then the Cesaro sum of the Fourier series, at $t = t_0$, exists and is equal to $\frac{1}{2}[f(t_{0^+}) + f(t_{0^-})]$. That is,

$$\lim_{n \to \infty} \sum_{j=-(n-1)}^{n-1} \left(1 - \frac{|j|}{n}\right) (f,\phi_j)\phi_j(t_0) = \frac{1}{2}[f(t_{0^+}) + f(t_{0^-})]$$

Furthermore, if f is continuous in a closed interval, the convergence is uniform for t_0 in that interval.

Proof We have seen that the nth average of the partial sums is precisely

$$\int_0^L K_n(t - t_0)f(t)\, dt$$

where

$$K_n(t) = \frac{1}{nL} \left[\frac{\sin (\pi nt/L)}{\sin (nt/L)}\right]^2$$

form a kernel system. Thus Theorem 1 applies.

The kernel system we have just described represents the *Fejer kernels*. The function $D_n(t)$ described above is called the *Dirichlet kernel*. This kernel must be studied for any deep investigation of convergence of a Fourier series; however, this is an investigation with which we shall not concern ourselves. It is an immediate consequence of the Fejer theorem that *if the Fourier series of a continuous function converges, it must converge to the function itself;* for we have seen that for a convergent series the Cesaro sum is the same as the value of the series.

Another immediate consequence of the Fejer theorem is the *uniqueness theorem*, Theorem 3.

Problem Using the function

$$f(t) = \frac{t(2\pi - t)}{\pi^2} \qquad 0 \le t \le 2\pi$$

show that

$$\sum_{n=1}^{\infty} \frac{1}{n^2} = \frac{\pi^2}{6} \qquad \sum_{n=1}^{\infty} \frac{(-1)^n}{n^2} = -\frac{\pi^2}{12}$$

Theorem 3 If f is a continuous function, of period L, and $(f, \phi_n) = 0$ for all n, then f is identically zero.

Thus two continuous functions with the same Fourier coefficients must be identical. In general, if two integrable functions have the same Fourier coefficients, at every point where they both have right- and left-hand limits the arithmetic mean of these two limits must be the same for both functions.

An important application of the Fejer theorem is the *Weierstrass approximation theorem*.

Theorem 4 Let f be a continuous function on a closed interval, $a \le t \le b$. Then for every $\epsilon > 0$ there exists a polynomial $p(t)$ such that $|f(t) - p(t)| < \epsilon$ for all t, $a \le t \le b$.

Proof Let $a' < a$ and $b' < b$. Consider the function $g(t)$ defined by

$$g(t) = \begin{cases} \dfrac{t - a'}{a - a'} f(a) & a' < t < a \\[2mm] \dfrac{t - a'}{a - a'} f(t) & a < t < b \\[2mm] \dfrac{t - b'}{b - b'} f(b) & b < t < b' \end{cases}$$

Here g is continuous for $a' \le t \le b'$, $g(a') = g(b') = 0$, and $g(t) = f(t)$ for $a \le t \le b$. Let $L = b' - a'$. Then the periodic extension of $g(t)$ is continuous. Given $\epsilon > 0$, from the Fejer theorem, there exists an n such that

$$\left| \sum_{j = -(n-1)}^{n-1} \left(1 - \frac{|j|}{n} \right) (g, \phi_j) \phi_j(t) - g(t) \right| < \frac{\epsilon}{2} \qquad \text{for } a \le t \le b$$

But

$$\sum_{j = -(n-1)}^{n-1} \left(1 - \frac{|j|}{n} \right) (g, \phi_j) \phi_j(t)$$

is a linear combination of exponential functions and has a power-series expansion which converges uniformly on every closed interval. Therefore there is a polynomial $p(t)$ such that

$$\left| p(t) - \sum_{j=-(n-1)}^{n-1} \left(1 - \frac{|j|}{n}\right)(g,\phi_j)\phi_j(t) \right| < \frac{\epsilon}{2} \qquad \text{for } a \leq t \leq b$$

Hence

$$|p(t) - f(t)| < \epsilon \qquad \text{for } a \leq t \leq b$$

THE POISSON KERNEL

Let $K(x,t)$ be, for each x, $0 \leq x < 1$, a function which satisfies the following conditions:

$K(x,t)$ is a continuous function of t, with period L, and
$K(x,-t) = K(x,t)$ (K-1′)

$K(x,t) \geq 0$ for all t (K-2′)

$$\int_0^L K(x,t)\, dt = 1 \qquad\qquad\qquad\qquad\qquad \text{(K-3′)}$$

and furthermore,

If $0 < t_1 < L/2$, then for every $\epsilon > 0$ there exists an x_0, $0 < x_0 < 1$, such that if $x_0 < x < 1$, then $K(x,t) < \epsilon$ for   (K-4′)
all t in the interval $t_1 \leq t \leq L/2$.

Then, with only slight changes, Theorem 1 remains true if

$$\lim_{n\to\infty} \int_0^L K_n(t - t_0)f(t)\, dt$$

is replaced by

$$\lim_{x\to 1} \int_0^L K(x, t - t_0)f(t)\, dt$$

We continue to call $K(x,t)$ a kernel system, and we shall now find a kernel system which plays the same role in the computation of the Abel sum of a Fourier series as did the Fejer kernels in the computation of the Cesaro sum.

Let f be an integrable function, of period L, with the Fourier coefficients (f,ϕ_n). In virtue of Bessel's inequality, the sequence $\{|f,\phi_n|\}$ is bounded, and thus the series

$$B(x,t_0) = (f,\phi_0)\phi_0(t_0) + \sum_{n=1}^{\infty} [(f,\phi_n)\phi_n(t_0) + (f,\phi_{-n})\phi_{-n}(t_0)]\, x^n$$

converges absolutely and uniformly on every interval $0 \leq x \leq x_1 < 1$. Hence $B(x,t_0)$ can be written as

$$B(x,t_0) = \int_0^L K(x, t - t_0)f(t)\, dt$$

where

$$K(x,t) = \frac{1}{L}\left\{1 + \sum_{n=1}^{\infty} [(xe^{2\pi it/L})^n + (xe^{-2\pi it/L})^n]\right\}$$

It is clear, since the series in this expression converges uniformly for each x, $0 \leq x < 1$, that $K(x,t)$ satisfies (K-1') and (K-3'). The series is geometric, and so it can be rewritten as

$$K(x,t) = \frac{1}{L}\frac{1 - x^2}{1 - 2x \cos (2\pi t/L) + x^2}$$

Since

$$1 - 2x \cos \frac{2\pi t}{L} + x^2 = \left(1 - x \cos \frac{2\pi t}{L}\right)^2 + x^2 \sin^2 \frac{2\pi t}{L}$$

$K(x,t) \geq 0$, and so (K-3') is satisfied. Furthermore, if $0 < t_1 \leq t \leq L/2$,

$$K(x,t) \leq \frac{1 - x^2}{L[1 - \cos (2\pi t/L)]^2}$$

satisfying (K-4').

Therefore we have the following theorem.

Theorem 5 Let f be integrable, with period L, and let t_0 be a point at which f has both right-hand and left-hand limits. Then the Abel sum of the Fourier series for f exists and is equal to $\frac{1}{2}[f(t_{0+}) + f(t_{0-})]$, and the limit

$$\lim_{x \to -1} \frac{1}{L} \int_0^L \frac{(1 - x^2)f(t)\, dt}{1 - 2x \cos (2\pi/L)(t - t_0) + x^2}$$

is uniform on every closed interval in which f is continuous.

Problem
(a) Show that

$$K(x,t) = \frac{1}{2\pi}\frac{1 - x^2}{1 - 2x \cos t + x^2}$$

satisfies the partial-differential equation

$$x\frac{\partial}{\partial x}\left(x\frac{\partial K}{\partial x}\right) + \frac{\partial^2 K}{\partial t^2} = 0$$

(b) Show that

$$P(r,\theta) = \frac{1}{2\pi} \frac{a^2 - r^2}{a^2 - 2ar \cos(\theta - \theta') + r^2}$$

satisfies the partial-differential equation

$$r^2 \frac{\partial^2 P}{\partial r^2} + r \frac{\partial P}{\partial r} + \frac{\partial^2 P}{\partial \theta^2} = 0 \qquad \text{for } r < a$$

(c) Show that the function

$$\psi(r,\theta) = \frac{1}{2\pi} \int_0^{2\pi} \frac{(a^2 - r^2)f(\theta')\,d\theta'}{a^2 - 2ar \cos(\theta - \theta') + r^2} \qquad \text{for } r < a$$

satisfies the partial-differential equation

$$\frac{\partial^2 \psi}{\partial r^2} + \frac{1}{r} \frac{\partial \psi}{\partial r} + \frac{1}{r^2} \frac{\partial^2 \psi}{\partial \theta^2} = 0$$

and that if $f(\theta)$ is integrable on $0 \le \theta \le 2\pi$, then

$$\lim_{r \to a^-} \psi(r,\theta) = \tfrac{1}{2}[f(\theta^+) + f(\theta^-)]$$

whenever this limit exists.

Problem

(a) Show that if $\psi(r,\theta)$ is a function of r and θ, $0 \le r < a$, $0 \le \theta \le 2\pi$, and if

$$\frac{\partial^2 \psi}{\partial r^2} + \frac{1}{r} \frac{\partial \psi}{\partial r} + \frac{1}{r^2} \frac{\partial^2 \psi}{\partial \theta^2} = g(r,\theta)$$

where $g(r,\theta)$ is integrable in both r and θ, then

$$\frac{d^2 \psi_n}{dr^2} + \frac{1}{r} \frac{d\psi_n}{dr} - \frac{n^2}{r^2} \psi_n = g_n(r)$$

where

$$f_n(r) = \frac{1}{2\pi} \int_0^{2\pi} f(r,\theta')e^{-in\theta'}\,d\theta'$$

for an integrable function $f(r,\theta)$.

(b) Show that for $\psi_n(r)$ to be a solution of the above equation which is bounded as $r \to 0$ and has the limit zero as $r \to a$, we must have

$$\psi_n(r) = \int_0^a G_n(r,r')r'g_n(r')\,dr'$$

where

$$G_0(r,r') = \begin{cases} \log \dfrac{r}{a} & \text{if } r' < r \\[2ex] \log \dfrac{r'}{a} & \text{if } r' > r \end{cases}$$

and

$$G_n(r,r') = \begin{cases} \dfrac{1}{2|n|}\left(\dfrac{rr'}{a^2}\right)^{|n|} - \left(\dfrac{r'}{r}\right)^{|n|} & \text{if } r' < r \\[3ex] \dfrac{1}{2|n|}\left(\dfrac{rr'}{a^2}\right)^{|n|} - \left(\dfrac{r}{r'}\right)^{|n|} & \text{if } r' > r \end{cases}$$

(c) Show that if $r' \neq r$, the series

$$\sum_{n=-\infty}^{\infty} G_n(r,r')e^{in(\theta-\theta')}$$

converges and is the Fourier series for the function

$$F(r,r',\theta,\theta') = \tfrac{1}{2} \log \frac{r^2 - 2rr' \cos(\theta' - \theta) + (r')^2}{a^2 - 2rr' \cos(\theta' - \theta) + (r'r)^2}$$

(d) Show that the Fourier series

$$\Sigma \psi_n(r)e^{in\theta}$$

converges and is the Fourier series of a function $\psi(r,\theta)$ which satisfies the above partial-differential equation, which is bounded as $r \to 0$ and has the limit zero as $r \to a$.

(e) Show that this function can be represented as

$$\psi(r,\theta) = \frac{1}{2\pi} \int_0^a \int_0^{2\pi} F(r,r',\theta,\theta')g(r',\theta')r' \, d\theta' \, dr'$$

4

Mean-square Convergence

THE MEAN-SQUARE LIMIT

If $\{f_N\}$ is a sequence of functions, integrable on $0 \leq t \leq L$, and if f is an integrable function such that

$$\lim_{N \to \infty} \int_0^L |f_N - f|^2 \, dt = 0$$

we say that f is the *mean-square limit* of f_N, and we adopt the notation

$$f = \text{m.s.l. } f_N$$

In case f_N is the Nth partial sum of an infinite series $\sum_n g_n$, we twist the English language a bit and write

$$f = \text{m.s.l. } \sum_n g_n$$

We saw in Chap. 1 that the necessary and sufficient condition that f, periodic and integrable, be the mean-square limit of its Fourier series is

that Parseval's equality be satisfied.   We shall prove in this chapter that
if f is continuous, Parseval's equality is satisfied.   It is true that this
result, along with its associated definitions and auxiliary results, can be
generalized to a much larger class of functions than the continuous
functions.   This extension would require a detailed discussion of the
theory of integration and the introduction of the Lebesgue integral,
which is beyond the scope of this text.   However, once we have proved
what we are about to prove for continuous functions, we shall see that it
is relatively easy to extend them to piecewise-continuous functions.

CONVOLUTION

Let f and g be continuous and periodic, of period L.   Then we shall
define the function $f * g$, the *convolution* of f and g, by the formula

$$(f * g)(t) = \int_0^L f(t - \tau)g(\tau)\, d\tau$$

It is obvious that $f * g$ is continuous and periodic, with period L.   If we
write

$$\int_0^L f(t - \tau)g(\tau)\, d\tau = - \int_t^{t-L} f(\zeta)g(t - \zeta)\, d\zeta$$

$$= \int_{t-L}^t g(t - \zeta)f(\zeta)\, d\zeta = \int_0^L g(t - \zeta)f(\zeta)\, d\zeta$$

we see that $f * g = g * f$.   The property of convolution which is of prime
importance for us is expressed in the following theorem.

Theorem 1

$$(f * g, \phi_n) = \sqrt{L}\,(f, \phi_n)(g, \phi_n)$$

Proof

$$\int_0^L \left[\int_0^L f(t - \tau)g(\tau)\, d\tau \right] \frac{e^{-2\pi n i t/L}}{\sqrt{L}}\, dt$$

$$= \int_0^L g(\tau) \left[\int_0^L \frac{f(t - \tau)e^{-2\pi n i t/L}}{\sqrt{L}}\, dt \right] d\tau$$

$$= \int_0^L g(\tau) \left[\int_{-\tau}^{L-\tau} f(t) \frac{e^{-2\pi n i t/L}}{\sqrt{L}}\, dt \right] e^{-2\pi n i \tau/L}\, d\tau$$

$$= (f, \phi_n) \int_0^L g(\tau)e^{-2\pi n i \tau/L}\, d\tau = \sqrt{L}\,(f, \phi_n)(g, \phi_n)$$

Theorem 2   If f^* is defined by the formula

$$f^*(t) = \bar{f}(-t)$$

then

$$(f^*, \phi_n) = \overline{(f, \phi_n)}$$

and

$$(f * f^*)(0) = \int_0^L |f|^2 \, dt$$

Proof

$$\int_0^L f^*(t) e^{-2\pi nit/L} \, dt = \int_0^L \bar{f}(-t) e^{-2\pi nit/L} \, dt$$

$$= \int_0^{-L} \bar{f}(t) e^{2\pi nit/L} \, dt = \overline{\int_0^L f(t) e^{-2\pi nit/L} \, dt}$$

$$\int_0^L f(t - \tau) f^*(\tau) \, d\tau = \int_0^L f(t - \tau) \bar{f}(-\tau) \, d\tau$$

$$= \int_0^L f(t + \tau) \bar{f}(\tau) \, d\tau$$

Theorem 3   If f is continuous and periodic, of period L, then

$$\sum_{n=-\infty}^{\infty} |(f, \phi_n)|^2 = \int_0^L |f|^2 \, dt$$

Proof   The function $f * f^*$ has the Fourier coefficients $\sqrt{L} \, |(f, \phi_n)|^2$. Therefore

$$f * f^* \sim \sum_{n=-\infty}^{\infty} \sqrt{L} \, |(f, \phi_n)|^2 \phi_n = \sum_{n=-\infty}^{\infty} |(f, \phi_n)|^2 e^{2\pi nit/L}$$

Because of Bessel's inequality, this Fourier series converges for all t, and since $f * f^*$ is continuous,

$$(f * f^*)(t) = \sum_{n=-\infty}^{\infty} |(f, \phi_n)|^2 e^{2\pi nit/L}$$

Setting $t = 0$ in this equality gives the assertion of the theorem.

This result can be improved considerably by only elementary methods. Let g be piecewise continuous; that is, let the interval $0 \le t \le L$ be subdivided into intervals

$$t_k \le t \le t_{k+1} \text{ and } 0 = t_0 < t_1 < \cdots < t_m = L$$

such that g has a right-hand limit at t_k and a left-hand limit at t_{k+1} and the function g_k, defined by

$$g_k(t) = \begin{cases} g(t_{k+}) & t = t_k \\ g(t) & t_k < t < t_{k+1} \\ g(t_{k+1-}) & t = t_{k+1} \end{cases}$$

is continuous on the closed interval $t_k \le t \le t_{k+1}$.   Then, if f is integrable,

$$\int_0^L f(t - \tau)g(\tau) \, d\tau = \sum_{k=0}^{m-1} \int_{t_k}^{t_{k+1}} f(t - \tau)g_k(\tau) \, d\tau$$

We wish to prove that

$$\int_0^L f(t - \tau)g(\tau) \, d\tau$$

is a continuous function of t.   It is clearly sufficient to prove that

$$\int_{t_k}^{t_{k+1}} f(t - \tau)g_k(\tau) \, d\tau$$

is a continuous function of t.   Let $\epsilon > 0$.   Then, if M is an upper bound for $|f|$, there exists a subdivision

$$t_k = t_{k,0} < t_{k,1} < \cdots < t_{k,n} = t_{k+1}$$

of the interval $t_k \le t \le t_{k+1}$ such that if $t_{k,j} \le t \le t_{k,j+1}, \ j = 0, 1, \ldots$
$n - 1$,

$$|g(t) - g(t_{k,j})| < \frac{\epsilon}{M(t_{k+1} - t_k)}$$

But then

$$\left| \int_{t_k}^{t_{k+1}} f(t - \tau)g_k(\tau) \, d\tau - \sum_{j=0}^{n-1} g(t_{k,j}) \int_{t_{k,j}}^{t_{k,j+1}} f(t - \tau) \, d\tau \right| < \epsilon$$

This shows that

$$\int_{t_k}^{t_{k+1}} f(t - \tau)g_k(\tau) \, d\tau$$

can be uniformly approximated (in t) by a function of the form

$$\sum_{j=0}^{n=1} g(t_{k,j}) \int_{t_{k,j}}^{t_{k,j+1}} f(t - \tau) \, d\tau$$

so that it is sufficient to prove that

$$\int_{t_{k,j}}^{t_{k,j+1}} f(t - \tau) \, d\tau$$

is a continuous function of t.   This integral is the same as

$$\int_{t-t_{k,j+1}}^{t-t_{k,j}} f(\tau) \, d\tau$$

which is clearly a continuous function of t.

Since Theorems 1 and 2 remain true for integrable functions, and Theorem 3 depends on these theorems and the fact that $f * f^*$ is continuous, we have established *Parseval's equality* for piecewise-continuous functions.

Theorem 4 If f is piecewise continuous and periodic, with period L,

$$f = \text{m.s.l.} \sum_{n=-\infty}^{\infty} (f,\phi_n)\phi_n$$

THE CAUCHY–SCHWARTZ INEQUALITY

We have already observed that (f,g) satisfies the following properties:

$$(f,f) \geq 0 \qquad\qquad\qquad\qquad\qquad\qquad\qquad\qquad\text{(C-1)}$$

$$(f,g) = (\bar{g},\bar{f}) \qquad\qquad\qquad\qquad\qquad\qquad\qquad\text{(C-2)}$$

$$(cf,g) = c(f,g) \qquad (f+h,\, g) = (f,g) + (h,g) \qquad\text{(C-3)}$$

As a simple consequence of these properties it is possible to prove the *Cauchy-Schwartz inequality*,

$$|(f,g)|^2 \leq (f,f)(g,g)$$

For any complex number λ,

$$0 \leq (f - \lambda g, f - \lambda g) = (f,f) - \lambda(g,f) - \bar{\lambda}(f,g) + |\lambda|^2(g,g)$$

Let $(f,g) = |(f,g)|e^{i\Phi}$, and set $\lambda = \mu e^{-i\Phi}$. Then for any real value of μ,

$$0 \leq (f,f) - 2\mu|(f,g)| + \mu^2(g,g)$$

If $(g,g) = 0$, this inequality is violated by taking μ sufficiently large, unless $|(f,g)| = 0$. Therefore if $(g,g) = 0$, then $(f,g) = 0$, and the equality is satisfied. If $(g,g) > 0$, let $\mu = |(f,g)|/(g,g)$. Then

$$0 \leq (f,f) - \frac{2|(f,g)|^2}{(g,g)} + \frac{|(f,g)|^2}{(g,g)} = (f,f) - \frac{|(f,g)|^2}{(g,g)}$$

which implies that

$$|(f,g)|^2 \leq (f,f)(g,g)$$

The fact that the Cauchy-Schwartz inequality follows from properties (C-1) to (C-3) has an immediate numerical consequence. Instead of letting f represent a function, let f represent a finite sequence of numbers,

$$a_1,\, a_2,\, \ldots,\, a_N$$

If g represents another finite sequence,

$$b_1,\, b_2,\, \ldots,\, b_N$$

let $f + g$ represent the sequence

$$a_1 + b_1,\, a_2 + b_2,\, \ldots,\, a_N + b_N$$

and for a constant c let cf represent the sequence

$$ca_1, \; ca_2, \; \ldots \; , \; ca_N$$

Now let us *define* (f,g) by the formula

$$(f,g) \;=\; \sum_{n=1}^{N} a_n \bar{b}_n$$

It is clear with this convention that (f,g) continues to satisfy (C-1) to (C-3). The Cauchy-Schwartz inequality then takes the following form: for two finite sequences of complex numbers $a_1, a_2, \ldots , a_N$ and $b_1, b_2, \ldots , b_N,$

$$\Big| \sum_{n=1}^{N} a_n \bar{b}_n \Big|^2 \;\leq\; \Big(\sum_{n=1}^{N} |a_n|^2 \Big) \sum_{n=1}^{N} |b_n|^2$$

Furthermore, since each a_n and each $\bar{b}_n$ in the left-hand side of this expression can be multiplied by a complex number of the form $e^{i\theta}$ without affecting the right-hand side, the inequality can be strengthened to

$$\sum_{n=1}^{N} |a_n|\,|b_n| \;\leq\; \Big(\sum_{n=1}^{N} |a_n|^2 \Big)^{1/2} \Big(\sum_{n=1}^{N} |b_n|^2 \Big)^{1/2}$$

With this result we are in a position to prove a final theorem regarding the convergence of Fourier series.

Theorem 5 Let f be a continuous function, of period L, and let f have a piecewise-continuous derivative. That is, let the interval $0 \leq t \leq L$ be subdivided into subintervals

$$0 = t_0 < t_1 < \; \cdots \; < t_k = L$$

in such a fashion that for $t_j < t < t_{j+1}$, $f' = df/dt$ exists and has a right-hand limit at t_j and a left-hand limit at t_{j+1}. Then the Fourier series for f converges absolutely and uniformly, and hence uniformly to f.

Proof　We first compute (f',ϕ_n):

$$(f',\phi_n) \;=\; \int_0^L \frac{df}{dt}\,\phi_n\,dt \;=\; \sum_{j=0}^{k-1} \int_{t_j}^{t_{j+1}} \phi_n \frac{df}{dt}\,dt$$

$$=\; \sum_{j=0}^{k-1} (\phi_n f) \Big]_{t_j}^{t_{j+1}} - \int_{t_j}^{t_{j+1}} f \frac{d\phi_n}{dt}\,dt$$

$$=\; \sum_{j=0}^{k-1} \phi_n f \Big]_{t_j}^{t_{j+1}} - \int_0^L f \frac{d\phi_n}{dt}\,dt$$

The sum in this last term vanishes, because $\phi_n f$ is continuous and of period L, so that

$$(f',\phi_n) = -\int_0^L f \frac{d\phi_n}{dt}\, dt = \frac{2\pi i n}{L}(f,\phi_n)$$

Therefore, if $n \neq 0$,

$$(f,\phi_n) = \frac{L}{2\pi i n}(f',\phi_n)$$

so that

$$\sum_{n=1}^N |(f,\phi_n)| + |f,\phi_{-n}|$$

$$= \frac{L}{2\pi} \sum_{n=1}^N \frac{1}{n} [|(f',\phi_n)| + |f',\phi_{-n}|]$$

$$\leq \frac{L}{2\pi} \left(\sum_{n=1}^N \frac{1}{n^2} \right)^{\!\frac12} \left\{ \left[\sum_{n=1}^N |(f',\phi_n)|^2 \right]^{\frac12} + \left[\sum_{n=1}^N |(f',\phi_n)|^2 \right]^{\frac12} \right\}$$

where the last inequality is a consequence of the Cauchy-Schwartz inequality for sequences. Therefore, because

$$\sum_{n=1}^\infty \frac{1}{n^2}$$

is a convergent series and the other two series,

$$\sum_{n=1}^\infty |(f',\phi_n)|^2 \qquad \text{and} \qquad \sum_{n=1}^\infty |(f',\phi_n)|^2$$

converge as a consequence of Bessel's inequality, the series

$$\sum_{n=1}^\infty |f,\phi_n| + |f,\phi_{-n}|$$

converges. Hence the series

$$\sum_{n=-\infty}^\infty (f,\phi_n)\phi_n$$

converges absolutely and uniformly.

5

The First-order Linear Equation with Constant Coefficients

FOURIER COEFFICIENTS OF PERIODIC SOLUTION

We shall investigate the differential equation

$$\frac{df}{dt} + af = F(t) \tag{E}$$

$F(t)$ is a piecewise-continuous function, of period L. f is a solution of (E) if f is continuous and has a piecewise-continuous derivative such that $df/dt + af = F(t)$ at all points at which $F(t)$ is continuous and $df/dt + af$ has the same left-hand and right-hand limits as $F(t)$ at the points at which $F(t)$ is discontinuous. If $F(t)$ is continuous, it is a fact learned in every course in differential equations that every solution of (E) is of the form

$$f(t) = e^{-at} \left[C + \int_0^t e^{a\tau} F(\tau) \, d\tau \right]$$

where C is a constant, and every function in this form is a solution.

Even when F is just piecewise continuous this expression provides us with a solution.

To solve (E) we seek a solution f which is periodic, of period L. Suppose that such a solution exists. Then, computing the Fourier coefficient of both sides of (E), we obtain

$$(f',\phi_n) + a(f,\phi_n) = (F,\phi_n)$$

But, from Chap. 4,

$$\left(\frac{2\pi i n}{L} + a\right)(f,\phi_n) = (F,\phi_n)$$

Thus if $a + 2\pi i n/L \neq 0$ for all integers n, the Fourier coefficients (f,ϕ_n) are determined by the formula

$$(f,\phi_n) = \frac{(F,\phi_n)}{2\pi i n/L + a}$$

Because we know that a continuous function is completely determined by its Fourier coefficients, the solution, if it exists, is unique. Conversely, if $a + 2\pi i n/L \neq 0$ for all n, then $e^{aL} - 1 \neq 0$, and the formula for $f(t)$ given above represents a periodic solution for

$$C = \frac{\int_0^L e^{a\tau} F(\tau)\, d\tau}{e^{aL} - 1}$$

Suppose now that for some integer n_0, $a + 2\pi i n_0/L = 0$; then, if $(F,\phi_{n_0}) \neq 0$, no solution exists. If, however, $(F,\phi_{n_0}) = 0$, the formula for $f(t)$ represents a periodic solution for all C. For such a solution the Fourier coefficients (f,ϕ_n), $n \neq n_0$, are determined as above, and (f,ϕ_{n_0}) is arbitrary.

We summarize these results in the following theorem.

Theorem 1 The differential equation (E) has a unique solution, of period L, if and only if $a + 2\pi n i/L \neq 0$ for all integers n. In this case the Fourier coefficients of the solution are given by the formula

$$(f,\phi_n) = \frac{(F,\phi_n)}{a + 2\pi n i/L}$$

If $a + 2\pi n i/L = 0$, for some integer n_0, then (E) has no solution unless $(F,\phi_{n_0}) = 0$. If $(F,\phi_{n_0}) = 0$, then (E) has solutions whose nth Fourier coefficient, $n \neq n_0$, is determined by the formula above and whose n_0th Fourier coefficient is arbitrary. The Fourier series of a solution converges uniformly to the solution.

It is interesting and useful to observe what happens in the case $F = \phi_{n_0}$ for some integer n_0. If a unique solution exists, $(f, \phi_n) = 0$ for $n \neq n_0$, and

$$f = \frac{1}{a + 2\pi n_0 i/L} \phi_{n_0}$$

This can be rewritten as

$$f = \frac{1}{\sqrt{a^2 + 4\pi^2 n_0^2/L^2}} e^{i \tan^{-1}(2\pi n_0/aL)} \phi_{n_0}$$

In a physical system in which f is a quantity determined by F by means of an equation of form (E), this last formula indicates that if F is a "pure" oscillation, the effect of the system is to amplify f by a factor

$$\frac{1}{\sqrt{a^2 + 4\pi^2 n_0^2/L^2}}$$

and induce a phase shift given by $\tan^{-1}(2\pi n_0/aL)$. As n_0 increases, the amplification factor goes to zero, and the phase shift goes to $\pm\pi/2$, depending upon whether a is positive or negative. Typical examples of such physical systems are series circuits containing a voltage supply, a resistance, and either an inductance or a capacitance.

Problem   Show that the solutions of the equation

$$\frac{df}{dt} + a(t)f = 0 \qquad a(t + L) = a(t)$$

are periodic, with period L, if and only if

$$\int_0^L a(t)\, dt = 2n\pi i \qquad n \text{ an integer}$$

HIGHER-ORDER EQUATIONS

Analogous results are easy to obtain for the equation

$$\frac{d^k f}{dt^k} + a_1 \frac{d^{k-1} f}{dt^{k-1}} + \cdots + a_k f = F(t) \tag{E_k}$$

If $F(t)$ is piecewise continuous, a solution of (E_k) must have its first $k - 1$ derivatives continuous and its kth derivative piecewise continuous. If $r_1, r_2, \ldots, r_k$ are the roots of the equation

$$r^k + a_1 r^{k-1} + \cdots + a_k = 0$$

then (E_k) is equivalent to the system of equations

$$\frac{df_1}{dt} - r_1 f_1 = F$$

$$\frac{df_2}{dt} - r_2 f_2 = f_1 \tag{E_k'}$$

$$\cdots\cdots\cdots\cdots$$

$$\frac{df}{dt} - r_k f = f_{k-1}$$

If none of the numbers $r_1, r_2, \ldots, r_k$ is of the form $2\pi ni/L$, for any integer n, the preceding section shows that we can successively find unique solutions $f_1, f_2, \ldots, f_{k-1}, f$ of the equations (E_k'), and the unique solution f of (E_k) has its Fourier coefficients determined by the formula

$$(f, \phi_n) = \frac{(F, \phi_n)}{(2\pi ni/L)^k + a_1 (2\pi ni/L)^{k-1} + \cdots + a_k}$$

The system of equations (E_k') is a special case of the system

$$\frac{df_1}{dt} + a_{11} f_1 + a_{12} f_2 + \cdots + a_{1k} f_k = F_1$$

$$\frac{df_2}{dt} + a_{21} f_1 + a_{22} f_2 + \cdots + a_{2k} f_k = F_2 \tag{S_k}$$

$$\cdots\cdots\cdots\cdots\cdots\cdots\cdots\cdots\cdots\cdots\cdots\cdots$$

$$\frac{df_k}{dt} + a_{k1} f_1 + a_{k2} f_2 + \cdots + a_{kk} f_k = F_k$$

The system (S_k) can be reduced to a system similar to (E_k') by algebraic devices equivalent to finding the Jordan canonical form for the matrix $[a_{ij}]$. If the functions $F_1, F_2, \ldots, F_k$ are piecewise continuous and of period L, then a unique set of functions $f_1, f_2, \ldots, f_k$, continuous, of period L, and with piecewise-continuous derivatives, exists which satisfies the system (S_k), provided that none of the characteristic values of the matrix $[a_{ij}]$ are of the form $2\pi ni/L$, for any integer n. In this case the Fourier coefficients of the solutions can be found by solving the system of linear equations

$$\left(\frac{2\pi ni}{L} + a_{11}\right)(f_1, \phi_n) + a_{12}(f_2, \phi_n) + \cdots + a_{1k}(f_k, \phi_n) = (F_1, \phi_n)$$

$$a_{21}(f_1, \phi_n) + \left(\frac{2\pi ni}{L} + a_{22}\right)(f_2, \phi_n) + \cdots + a_{2k}(f_k, \phi_n) = (F_2, \phi_n)$$

$$\cdots\cdots\cdots\cdots\cdots\cdots\cdots\cdots\cdots\cdots\cdots\cdots\cdots$$

$$a_{k1}(f_1, \phi_n) + a_{k2}(f_2, \phi_n) + \cdots + \left(\frac{2\pi ni}{L} + a_{kk}\right)(f_k, \phi_n) = (F_k, \phi_n)$$

Problem   Find the solutions with period 2π of the following differential equations:

(a) $\dfrac{d^2f}{dt^2} - 2\dfrac{df}{dt} + 2f = \cos 2t$

(b) $\dfrac{d^3f}{dt^3} - 4\dfrac{d^2f}{dt^2} + 5\dfrac{df}{dt} - 2f = \sin t$

Answers

(a) $f = -\frac{1}{26}(3\cos 2t + 2\sin 2t)$
(b) $f = \frac{1}{10}(\sin t - 2\cos t)$

Problem   Find the solutions with period 2π of the system

$$\frac{dx}{dt} - 2x + y = 3\sin t$$

$$\frac{dy}{dt} + x - y = 0$$

Answer

$x = -(\cos t + \sin t)$
$y = -(\cos t)$

6
The Fourier Integral

THE FOURIER TRANSFORM

Let f be defined for all values of t and integrable on every finite interval, and let the improper integral

$$\int_{-\infty}^{\infty} |f| \, dt$$

exist. Then the function $\hat{f}$, defined by the formula

$$\hat{f}(\omega) = \frac{1}{\sqrt{2\pi}} \int_{-\infty}^{\infty} f(t) e^{-i\omega t} \, dt = \lim_{N \to \infty} \frac{1}{\sqrt{2\pi}} \int_{-N}^{N} f(t) e^{-i\omega t} \, dt$$

exists and is called the *Fourier transform* of f. The function $\hat{f}$ bears an analogy to the Fourier coefficient, and to extend the analogy to the Fourier series, we define the *Fourier integral*,

$$\frac{1}{\sqrt{2\pi}} \int_{-\infty}^{\infty} \hat{f}(\omega) e^{i\omega t} \, d\omega$$

Let us now study the question of convergence of the Fourier integral.

Problem   Find the Fourier transform of the following functions:

$$(a)\ f(t) = \begin{cases} \dfrac{1}{T} & -\dfrac{T}{2} < t < \dfrac{T}{2} \\[2mm] 0 & |t| \geq \dfrac{T}{2} \end{cases}$$

$(b)\ f(t) = e^{-|t|}$

$(c)\ f(t) = e^{-t^2/2a^2}$

Answers

$(a)\quad \dfrac{1}{\sqrt{2\pi}}\ \dfrac{\sin(\omega T/2)}{\omega T/2}$

$(b)\quad \sqrt{\dfrac{2}{\pi}}\ \dfrac{1}{1 + \omega^2}$

$(c)\quad ae^{-a^2\omega^2/2}$

CESARO SUMMATION OF IMPROPER INTEGRALS

Given a function A which is integrable on every finite interval, it may be that the improper integral

$$\int_{-\infty}^{\infty} A(\omega)\ d\omega = \lim_{r \to \infty} \int_{-r}^{r} A(\omega)\ d\omega$$

does not exist. However, if we compute

$$\lim_{N \to \infty} \frac{1}{N} \int_0^N \left[\int_{-r}^{r} A(\omega)\ d\omega \right] dr$$

this limit may exist. If it does, it is called the *Cesaro sum* of the integral $\int_{-\infty}^{\infty} A(\omega)\ d\omega$, which is commonly abbreviated as

$$\int_{-\infty}^{\infty} A(\omega)\ d\omega = \lim_{N \to \infty} \frac{1}{N} \int_0^N \left[\int_{-r}^{r} A(\omega)\ d\omega \right] dr$$

$$= \lim_{N \to \infty} \int_{-N}^{N} \left(1 - \frac{|\omega|}{N} \right) A(\omega)\ d\omega \qquad \text{(C-1)}$$

Theorem 1   If $\int_{-\infty}^{\infty} A(\omega)\ d\omega$ exists, then so does the Cesaro sum of $\int_{-\infty}^{\infty} A(\omega)\ d\omega$, and the two numbers are equal.

Proof   The proof is similar to that for the corresponding theorem for series.   If

$$\int_{-\infty}^{\infty} A(\omega)\, d\omega = a$$

then for $\epsilon > 0$ there exists an r_0 such that if $r > r_0$,

$$\left| \int_{-r}^{r} A(\omega)\, d\omega - a \right| < \frac{\epsilon}{2}$$

Then

$$\frac{1}{r_0 + X} \int_{0}^{r_0 + X} \int_{-r}^{r} A(\omega)\, d\omega\, dr - a$$

$$= \frac{1}{r_0 + X} \left[\int_{0}^{r_0} \int_{-r}^{r} A(\omega)\, d\omega\, dr - r_0 a \right]$$

$$+ \frac{1}{r_0 + X} \int_{r_0}^{r_0 + X} \left[\int_{-r}^{r} A(\omega)\, d\omega - a \right] dr$$

In the right-hand side of the last equality, the second term is less in absolute value than $\epsilon/2$.   By taking X sufficiently large, the first term can be made less in absolute value than $\epsilon/2$.

ABEL SUMMATION OF INTEGRALS

Let g be a function, continuous for $x > 0$.   The improper integral

$$\int_{0}^{\infty} g(x)\, dx$$

may not exist.   However, if for all $\alpha > 0$ the integral

$$\int_{0}^{\infty} g(x) e^{-\alpha x}\, dx$$

exists, and if

$$\lim_{\alpha \to 0} \int_{0}^{\infty} g(x) e^{-\alpha x}\, dx$$

exists, then this limit is called the *Abel sum* of the integral $\int_{0}^{\infty} g(x)\, dx$.

Theorem 2   If $g(x)$ is continuous and bounded, and if $\int_{0}^{\infty} g(x)\, dx$ exists, then its Abel sum exists, and the two numbers are equal.

Proof   If $\epsilon > 0$, there exists an X_0 such that if $X > X_0$,

$$\left| \int_{X_0}^{X} g(x)\, dx \right| < \frac{\epsilon}{4}$$

Let $X > X_0$; then

$$\int_0^X g(x)\, dx - \int_0^X g(x)e^{-\alpha x}\, dx = \int_0^{X_0} g(x)(1 - e^{-\alpha x})\, dx$$
$$+ (1 - e^{-\alpha x})\int_{X_0}^X g(x)\, dx + \alpha \int_{X_0}^X \left[\int_{X_0}^X g(\zeta)\, d\zeta\right] e^{-\alpha x}\, dx$$

Thus

$$\left|\int_0^X g(x)\, dx - \int_0^X g(x)e^{-\alpha x}\, dx\right| \leq \left|\int_0^{X_0} g(x)(1 - e^{-\alpha x})\, dx\right|$$
$$+ (1 - e^{-\alpha X})\frac{\epsilon}{4} + (e^{-\alpha X_0} - e^{-\alpha X})\frac{\epsilon}{4} \qquad \text{for all } X > X_0$$

and hence

$$\left|\int_0^\infty g(x)\, dx - \int_0^\infty g(x)e^{-\alpha x}\, dx\right| \leq \int_0^{X_0}\left|g(x)\right|(1 - e^{-\alpha x})\, dx + \frac{\epsilon}{2}$$

If α is sufficiently small, the integral on the right can be made less than $\epsilon/2$.

An immediate application of Theorem 2 is in the evaluation of

$$\int_0^\infty \frac{\sin x}{x}\, dx$$

which can be shown to converge if we write it as

$$\sum_{j=0}^\infty \int_{j\pi}^{(j+1)\pi} \frac{\sin x\, dx}{x}$$

and observe that this series satisfies the well-known criteria for the convergence of an alternating series. The function

$$F(\alpha) = \int_0^\infty \frac{\sin x}{x} e^{-\alpha x}\, dx$$

must then be evaluated and its limit computed as $\alpha \to 0$ to find the value of the original integral. But

$$F'(\alpha) = -\int_0^\infty \sin x\, e^{-\alpha x}\, dx = -\frac{1}{1 + \alpha^2}$$

and therefore

$$F(\alpha) = C - \tan^{-1}\alpha$$

Since $F(\alpha) \to 0$ as $\alpha \to \infty$, $C = \pi/2$, and so $\lim_{\alpha \to 0} F(\alpha) = \pi/2$.

THE CESARO SUM OF THE FOURIER INTEGRAL

We can now prove the theorem for Fourier integrals which is analogous to the Fejer theorem for Fourier series.

Theorem 3 Let f be integrable on every finite interval, and let

$$\int_{-\infty}^{\infty} |f| \, dt$$

exist. If t_0 is a point at which f has both right-hand and left-hand limits, and if $\hat{f}$ is the Fourier transform of f, then the Cesaro sum of

$$\frac{1}{\sqrt{2\pi}} \int_{-\infty}^{\infty} \hat{f}(\omega) e^{i\omega t_0} \, d\omega$$

is equal to

$$\tfrac{1}{2}[f(t_{0^+}) + f(t_{0^-})]$$

Proof

$$\frac{1}{\sqrt{2\pi}} \int_{-r}^{r} \hat{f}(\omega) e^{i\omega t_0} \, d\omega = \frac{1}{2\pi} \int_{-r}^{r} \int_{-\infty}^{\infty} f(t) e^{i\omega(t_0 - t)} \, dt \, d\omega$$

$$= \int_{-\infty}^{\infty} f(t) D(r; t - t_0) \, dt$$

where

$$D(r;t) = \frac{1}{\pi} \frac{\sin rt}{t}$$

We observe that for $r > 0$

$$\int_{-\infty}^{\infty} D(r;t) \, dt = \frac{2}{\pi} \int_{0}^{\infty} \frac{\sin x}{x} \, dx = 1$$

Then

$$\frac{1}{N} \int_{0}^{N} \left[\frac{1}{\sqrt{2\pi}} \int_{-r}^{r} \hat{f}(\omega) e^{i\omega t_0} \, d\omega \right] dr = \frac{1}{N} \int_{0}^{N} \left[\int_{-\infty}^{\infty} f(t) D(r; t - t_0) \, dt \right] dr$$

$$= \int_{-\infty}^{\infty} K(N, t - t_0) f(t) \, dt$$

and

$$K(N,t) = \frac{1}{N} \int_{0}^{N} D(r;t) \, dr = \frac{2}{N\pi} \left[\frac{\sin (Nt/2)}{t} \right]^2$$

$K(N,t)$ has the following properties:

$K(N,t)$ is continuous in t, and $K_n(-t) = K_n(t)$ (K-1″)

$K(N,t) \geq 0$ for all t (K-2″)

$$\int_{-\infty}^{\infty} K(N,t) \, dt = 1$$ (K-3″)

If $t \geq t_1 > 0$, then for $\epsilon > 0$ there exists an N_0 such that if $N > N_0$, $|K(N,t)| < \epsilon$ (K-4″)

The proof that

$$\lim_{N\to\infty} \int_{-\infty}^{\infty} K(N, t - t_0)f(t)\, dt = \tfrac{1}{2}[f(t_{0^+}) + f(t_{0^-})]$$

is then, with minor modifications, the same as the proofs in Chap. 3.

We can now draw conclusions similar to those for the case of Fourier series: if f is integrable on every finite interval, and if

$$\int_{-\infty}^{\infty} |f|\, dt$$

exists, then at every point at which the Fourier integral of f exists, it must be equal to $\tfrac{1}{2}[f(t_{0^+}) + f(t_{0^-})]$, if this last expression has meaning. A continuous function is completely determined by its Fourier transform and is the Fourier integral of that transform when the integral exists.

THE PARSEVAL THEOREMS

We shall now prove for the Fourier integral, the theorems involving mean-square convergence which are analogous to those theorems we proved for the Fourier series. Because we do not have any obvious relationship analogous to $(\phi_j, \phi_k) = \delta_{jk}$, our approach will be somewhat different, although it could also have been applied to the earlier problem. We shall, for simplicity, restrict the discussion to continuous functions. On an interval $-N \leq t \leq N$ we define

$$(f,g)_N = \int_{-N}^{N} f(t)\bar{g}(t)\, dt$$

Then, as before, we have the inequality

$$|(f,g)_N| \leq (f,f)_N (g,g)_N$$

which implies, as for sequences,

$$\int_{-N}^{N} |f|\, |g|\, dt \leq \left(\int_{-N}^{N} |f|^2\, dt\right)^{\frac{1}{2}} \left(\int_{-N}^{N} |g|^2\, dt\right)^{\frac{1}{2}}$$

Thus if f and g are continuous functions for which the integrals

$$\int_{-\infty}^{\infty} |f|^2\, dt \qquad \text{and} \qquad \int_{-\infty}^{\infty} |g|^2\, dt$$

both exist, then

$$\int_{-\infty}^{\infty} f(t)\bar{g}(t)\, dt$$

exists and is, in fact, absolutely convergent. We define for such functions

$$(f,g) = \int_{-\infty}^{\infty} f(t)\bar{g}(t)\, dt$$

In the remainder of this chapter, every function f considered will be continuous, and

$$\int_{-\infty}^{\infty} |f|^2 \, dt \qquad \text{and} \qquad \int_{-\infty}^{\infty} |f| \, dt$$

will exist.

If f and g are given, then

$$\int_{-N}^{N} f(t-\tau)g(\tau) \, d\tau$$

exists for every N. Furthermore,

$$\int_{-N}^{N} |f(t-\tau)| \, |g(\tau)| \, d\tau \leq (f,f)^{1/2}(g,g)^{1/2}$$

so that the function $f * g$, defined by the formula

$$(f * g)(t) = \int_{-\infty}^{\infty} f(t-\tau)g(\tau) \, d\tau$$

exists, with the improper integral converging uniformly in t. Hence $f * g$ is a continuous function of t. Furthermore,

$$\int_{-N}^{N} |f * g| \, dt \leq \int_{-N}^{N} \int_{-\infty}^{\infty} |f(t-\tau)|g(\tau)| \, d\tau \, dt \leq \int_{-\infty}^{\infty} |g| \, dt \int_{-\infty}^{\infty} |f| \, dt$$

Therefore $f * g$ has a Fourier transform $\widehat{f * g}$, and as in the case of Fourier coefficients,

$$\widehat{f * g} = \sqrt{2\pi} \, \hat{f}\hat{g}$$

In particular, if f^* is defined by the formula

$$f^*(t) = \bar{f}(-t)$$

we have, as in the case of periodic functions,

$$\hat{f}^* = \bar{\hat{f}} \qquad (f * f^*)(0) = (f,f)$$

so that $f * f$ has the Fourier transform $\sqrt{2\pi} \, |\hat{f}|^2$. Thus $(f * f)(t)$ is the Cesaro sum of the Fourier integral of $\sqrt{2\pi} \, |\hat{f}|^2$, and in particular, (f,f) is equal to the Cesaro sum of

$$\int_{-\infty}^{\infty} |\hat{f}|^2 \, d\omega$$

We now state the theorem known as *Parseval's equality*.

Theorem 4 If f is continuous, and if both

$$\int_{-\infty}^{\infty} |f| \, dt \qquad \text{and} \qquad \int_{-\infty}^{\infty} |f|^2 \, dt$$

exist, then

$$\int_{-\infty}^{\infty} |f|^2 \, dt = \int_{-\infty}^{\infty} |\bar{f}|^2 \, d\omega$$

Proof We have established in the preceding remarks that

$$\int_{-\infty}^{\infty} |f|^2 \, dt$$

equals the Cesaro sum of

$$\int_{-\infty}^{\infty} |\bar{f}|^2 \, d\omega$$

We shall have proved the theorem if we can establish that if $A(\omega) \geq 0$, and if the Cesaro sum of

$$\int_{-\infty}^{\infty} A(\omega) \, d\omega$$

exists, then

$$\int_{-\infty}^{\infty} A(\omega) \, d\omega$$

exists. To show this, we write the Cesaro sum of

$$\int_{-\infty}^{\infty} A(\omega) \, d\omega$$

as

$$\lim_{n \to \infty} \int_{-N}^{N} \left(1 - \frac{|\omega|}{N}\right) A(\omega) \, d\omega = \lim_{N \to \infty} \int_{0}^{N} \left(1 - \frac{\omega}{N}\right) B(\omega) \, d\omega$$

where $B(\omega) = A(\omega) + A(-\omega) \geq 0$. If we let $\epsilon > 0$, then for some $N_0 > 0$, $N > N_0$, and $X > 0$,

$$\left| \int_{0}^{N+X} \left(1 - \frac{\omega}{N+X}\right) B(\omega) \, d\omega - \int_{0}^{N} \left(1 - \frac{\omega}{N}\right) B(\omega) \, d\omega \right| < \epsilon$$

But

$$\int_{0}^{N+X} \left(1 - \frac{\omega}{N+X}\right) B(\omega) \, d\omega - \int_{0}^{N} \left(1 - \frac{\omega}{N}\right) B(\omega) \, d\omega$$

$$= \int_{0}^{N} \frac{X}{N+X} \frac{\omega}{N} B(\omega) \, d\omega + \int_{N}^{N+X} \left(1 - \frac{\omega}{N+X}\right) B(\omega) \, d\omega \geq 0$$

Therefore, for $N > N_0$ and $X > 0$,

$$0 \leq \int_{0}^{N} \frac{X}{N+X} \frac{\omega}{N} B(\omega) \, d\omega < \epsilon$$

And if we let X be arbitrarily large,

$$0 \leq \int_0^N \frac{\omega}{N} B(\omega) \, d\omega \leq \epsilon \qquad \text{for } N > N_0$$

This implies that

$$\lim_{N \to \infty} \int_0^N B(\omega) \, d\omega$$

exists and is the same as

$$\lim_{N \to \infty} \int_0^N \left(1 - \frac{\omega}{N} \right) B(\omega) \, d\omega$$

But $\int_0^N B(\omega) \, d\omega = \int_{-N}^N A(\omega) \, d\omega$

and so the theorem is proved.

This theorem can easily be extended, as in the periodic case, to f piecewise continuous on every interval. The following *Parseval theorem* is stated without proof.

Theorem 5   If f satisfies the hypotheses of Theorem 4, then

$$\lim_{r \to \infty} \int_{-\infty}^{\infty} \left| f(t) - \frac{1}{\sqrt{2\pi}} \int_{-r}^{r} \hat{f}(\omega) e^{i\omega t} \, d\omega \right|^2 dt = 0$$

Theorem 5 is the counterpart for Fourier integrals of the fact that for periodic functions the function is the mean-square limit of its Fourier series.   Proof of Theorem 5 would carry us beyond the limits of this text. It is sufficient, for many applications, to know that if the Fourier integral for f converges, it converges to f.

If f is continuous and has a derivative f' which is piecewise continuous, and if both

$$\int_{-\infty}^{\infty} |f| \, dt \qquad \text{and} \qquad \int_{-\infty}^{\infty} |f'| \, dt$$

exist, then

$$\hat{f}' = i\omega \hat{f}$$

If, in addition,

$$\int_{-\infty}^{\infty} |f'|^2 \, d\omega$$

exists, Parseval's equality, Theorem 4, shows us that

$$\int_{-\infty}^{\infty} |\hat{f}'|^2 \, d\omega$$

exists, and therefore, for either $0 < r_1 < r_2$ or $r_1 < r_2 < 0$,

$$\int_{r_1}^{r_2} |\hat{f}| \, d\omega \leq \left(\int_{r_1}^{r_2} \frac{1}{\omega^2} \, d\omega \right)^{\frac{1}{2}} \left(\int_{r_1}^{r_2} |\hat{f}'|^2 \, d\omega \right)^{\frac{1}{2}}$$

so that the Fourier integral for f converges uniformly to f.

Problem

(a) Let $f(t)$ be differentiable, and let

$$\int_{-\infty}^{\infty} |f(t)|^2 \, dt \qquad \text{and} \qquad \int_{-\infty}^{\infty} |f'(t)|^2 \, dt$$

exist. Show that

$$\hat{f}' = i\omega \hat{f}$$

and that

$$\int_{-\infty}^{\infty} \omega^2 |\hat{f}|^2 \, d\omega = \int_{-\infty}^{\infty} |f'|^2 \, dt$$

(b) If

$$\int_{-\infty}^{\infty} t^2 |f|^2 \, dt$$

exists, prove the inequality

$$\int_{-\infty}^{\infty} |f|^2 \, dt \int_{-\infty}^{\infty} |\hat{f}|^2 \, d\omega \leq 4 \int_{-\infty}^{\infty} t^2 |f|^2 \, dt \int_{-\infty}^{\infty} \omega^2 |\hat{f}|^2 \, d\omega$$

(c) Show that the equality holds if $f = e^{-t^2/2a^2}$.

Problem Let $f(x,y)$ be a function for which

$$\frac{\partial^2 f}{\partial x^2} + \frac{\partial^2 f}{\partial y^2} = 0 \qquad \text{for } -\infty < x < \infty, \, y \geq 0$$

Suppose that

$$f(x,0) = g(x) \qquad \text{and} \qquad \lim_{y \to \infty} f(x,y) = 0 \qquad \text{for all } x$$

Suppose further that

$$\int_{-\infty}^{\infty} |f(x,y)|^2 \, dx$$

exists for all y.

(a) Let $\hat{f}(\omega,y)$ be the Fourier transform of $f(x,y)$, with y regarded as a constant. Show that

$$\hat{f}(\omega,y) = \hat{g}(\omega)e^{-|\omega|y}$$

(b) Show that $e^{-|\omega|y}$ is the Fourier transform of

$$\sqrt{\frac{2}{\pi}} \frac{y}{x^2 + y^2}$$

(c) Show that

$$f(x,y) = \frac{1}{\pi} \int_{-\infty}^{\infty} \frac{y}{y^2 + (x - x')^2} g(x') \, dx'$$

Problem   Suppose that $f(x,t)$ is a function for which

$$\frac{\partial^2 f}{\partial x^2} - \frac{\partial^2 f}{\partial t^2} = 0 \qquad -\infty < x < \infty, t \geq 0$$

and that

$$f(x,0) = g(x) \qquad \text{and} \qquad \frac{\partial f}{\partial t}(x,0) = h(x)$$

Suppose also that

$$\int_{-\infty}^{\infty} |f(x,t)|^2 \, dx \qquad \text{and} \qquad \int_{-\infty}^{\infty} \left| \frac{\partial f(x,t)}{\partial t} \right|^2 dx$$

exist for all t.   Let $\hat{f}(\omega,t)$ be the Fourier transform of $f(x,t)$, with t regarded as a constant.
(a) Show that

$$\hat{f}(\omega,t) = \hat{g}(\omega) \cos \omega t + \hat{h}(\omega) \frac{\sin \omega t}{\omega}$$

(b) Show that

$$f(x,t) = \tfrac{1}{2}[g(x - t) + g(x + t)] + \tfrac{1}{2} \int_0^t [h(x - \tau) + h(x + \tau)] \, d\tau$$

Problem   Suppose that $f(x,t)$ is a function for which

$$\frac{\partial f}{\partial t} = k^2 \frac{\partial^2 f}{\partial x^2} \qquad -\infty < x < \infty, t \geq 0$$

and that

$$f(x,0) = g(x) \qquad \text{and} \qquad \int_{-\infty}^{\infty} |f|^2 \, dx = 0$$

Let $\hat{f}(\omega,t)$ be the Fourier transform of 4.
(a) Show that

$$\hat{f}(\omega,t) = \hat{g}(\omega)e^{-k^2\omega^2 t}$$

(b) Show that

$$f(x,t) = \frac{1}{2k \sqrt{\pi t}} \int_{-\infty}^{\infty} g(x')e^{-(x-x')^2/4k^2 t} \, dx'$$

7

Generalized Fourier Series

ORTHONORMAL SEQUENCE

Suppose that for a class of functions, defined on an interval of real
numbers, a number (f,g) is defined for each pair of functions in the class.
Suppose the class of functions is such that if f and g are in the class, then
so is $af + bg$ for every pair of constants a and b. Then, if (f,g) satisfies
the conditions

$$(f,f) \geq 0$$
$$(f,g) = \overline{(g,f)}$$
$$(af + bg, h) = a(f,h) + b(g,h)$$

the number (f,g) is the *inner product* for the class of functions. If $\{\psi_n\}$,
$n = 0, 1, 2, \ldots$, is a sequence of functions in the class which satisfy
the equalities

$$(\psi_j, \psi_k) = \delta_{jk}$$

then $\{\psi_n\}$ is said to be an *orthonormal sequence*.

If $\{\psi_n\}$ is an orthonormal sequence and f is in the class, the numbers (f,ψ_n) are called the *generalized Fourier coefficients* of f; similarly, the series

$$\sum_{n=0}^{\infty} (f,\psi_n)\psi_n$$

is called the *generalized Fourier series* for f. If, in the periodic case, we define

$$\psi_{2p} = \phi_{-p} \qquad \psi_{2p+1} = \phi_{p+1} \qquad p = 0, 1, 2, \ldots$$

we see that the Fourier series we have already studied are a special case of the generalized Fourier series. We introduce the notation

$$\|f\| = (f,f)^{1/2}$$

The number $\|f\|$ is called the *norm* of f, and the Cauchy-Schwartz inequality is

$$|(f,g)| \le \|f\| \, \|g\|$$

As in the periodic case, using just the properties of the inner product and the definitions above, we can prove the following theorem.

Theorem 1   Let $\{\psi_n\}$ be an orthonormal sequence in a class of functions, and let f be in the class. Then, for a fixed integer N, among all sets of constants $(a_0, a_1, \ldots, a_n)$, that set which minimizes the expression

$$\left\| f - \sum_{n=0}^{N} a_n\psi_n \right\|$$

is the set

$$a_n = (f,\psi_n) \qquad n = 0, 1, \ldots, N$$

The generalized Fourier coefficients satisfy Bessel's inequality,

$$\sum_{n=0}^{\infty} |(f,\psi_n)|^2 < \|f\|^2$$

and a necessary and sufficient condition that

$$\lim_{N\to\infty} \left\| f - \sum_{n=0}^{N} (f,\psi_n)\psi_n \right\| = 0$$

is Parseval's equality,

$$\sum_{n=0}^{\infty} |(f,\psi_n)|^2 = \|f\|^2$$

If Parseval's equality is satisfied for every f in the class of functions, the orthonormal sequence $\{\psi_n\}$ is said to be *complete* in the class. It is clear that if $\{\psi_n\}$ is complete, then

$$(f,\psi_n) = 0 \qquad \text{for all } n$$

implies that

$$\|f\| = 0$$

However, if $(f,\psi_n) = 0$ for all n implies that $\|f\| = 0$, we can prove Parseval's equality only by making an additional hypothesis about the class of functions: if $\{f_n\}$ is a sequence of functions in the class, and if for every $\epsilon > 0$ there exists an N such that for $n > N$ and $m > N$,

$$\|f_n - f_m\| < \epsilon$$

then there exists a function f in the class such that

$$\lim_{n \to \infty} \|f - f_n\| = 0$$

Such a class of functions is called a *complete class* of functions. Suppose, for example, our class of functions were the continuous functions on the interval $0 < t < 1$, and we let

$$(f,g) = \int_a^1 f\bar{g}\, dt$$

It is easy to see that such a class of functions would not be complete, for the sequence $\{f_n\}$ of continuous functions would converge monotonically to the function f, defined as

$$f(t) = \begin{cases} 0 & \text{for } 0 < t < \frac{1}{2} \\ 1 & \text{for } \frac{1}{2} \le t \le 1 \end{cases}$$

THE LEGENDRE POLYNOMIALS

Let our class of functions be those which are continuous on the interval $-1 \le x \le 1$. If f and g are in the class, let

$$(f,g) = \int_{-1}^1 f\bar{g}\, dx$$

We shall define a certain sequence of polynomials, which, when modified slightly, become a complete orthonormal sequence. These functions arise in a natural fashion from the consideration of the function $1/r$, where r is the distance between two points in space. If p and p_0 are, respectively, the distances of the two points from the origin, and γ is the angle between the lines connecting the two points to the origin, then the law of cosines states that

$$r^2 = p^2 + p_0{}^2 - 2pp_0 \cos \gamma$$

Thus, if we assume that $p > p_0$,

$$\frac{1}{r} = \frac{1}{p} \frac{1}{\sqrt{1 - 2p_0/p \cos \gamma + (p_0/p)^2}}$$

or

$$\frac{1}{r} = \frac{1}{p} \frac{1}{\sqrt{1 - 2zx + z^2}}$$

where $z = p_0/p$ and $x = \cos \gamma$.

Now consider, for x fixed, the quadratic form

$$1 - 2xz + z^2$$

The two roots of this form are $x \pm i \sqrt{1 - x^2}$, so that as long as $-1 < x < 1$ and $|z| < 1$, this form is not zero. Hence the function

$$(1 - 2xz + z^2)^{-\frac{1}{2}}$$

has, for $z < 1$, a power-series representation

$$(1 - 2xz + z^2)^{-\frac{1}{2}} = \sum_{n=0}^{\infty} P_n(x)z^n \qquad (L)$$

The $P_n(x)$ are given as

$$\frac{1}{n!} \left(\frac{d}{dz} \right)^n (1 - 2xz + z^2)^{-\frac{1}{2}} \bigg]_{z=0}$$

and are clearly polynomials in x. These polynomials are called the *Legendre polynomials*.

Theorem 2   The Legendre polynomial $P_n(x)$ is a polynomial of degree n. Only even or only odd powers of x will appear in $P_n(x)$, according to whether n is even or odd. $P_n(x)$ satisfies the differential equation

$$\frac{d}{dx} \left[(1 - x^2) \frac{dP_n}{dx} \right] + n(n + 1)P_n(x) = 0$$

The functions $\psi_n(x) = \sqrt{n + \frac{1}{2}}\, P_n(x)$ form a complete orthonormal system.

Proof   The polynomials $P_n(x)$ can be computed explicitly by expanding the left member of (L) as a power series. However, there is a more efficient way to proceed. First, we observe that if the left member of (L) is differentiated with respect to x, we obtain

$$z(1 - 2xz + z^2)^{-3/2} = \sum_{n=0}^{\infty} \frac{dP_n(x)}{dx} z^n \qquad \text{(L')}$$

Therefore, letting $x = 0$ in both (L) and (L'), we have

$$(1 + z^2)^{-1/2} = \sum_{n=0}^{\infty} P_n(0) z^n \qquad z(1 + z^2)^{-3/2} = \sum_{n=0}^{\infty} \frac{dP_n(0)}{dx} z^n$$

From these equalities it is apparent that $P_n(0) = 0$ if n is odd, and $dP_n(0)/dx$ can be found by using the binomial expansion of the left-hand members of these equalities. For n even, this statement remains true when $P_n(0)$ and $dP_n(0)/dx$ are interchanged. Differentiating (L') once again with respect to x, we have

$$(1 - 2xz + z^2)^{-3/2} + 3z^2(1 - 2xz + z^2)^{-5/2} = \sum_{n=0}^{\infty} \frac{d^2 P_n(x)}{dx} z^n \qquad \text{(L'')}$$

Now, multiplying both members of (L) by z and differentiating twice with respect to z, we obtain

$$z \frac{\partial^2}{\partial z^2} [z(1 - 2xz + z^2)^{-1/2}] = \sum_{n=0}^{\infty} n(n + 1) P_n(x) z^n$$

This equality, when used with (L') and (L''), gives

$$\sum_{n=0}^{\infty} \left[(1 - x^2) \frac{d^2 P_n}{dx^2} - 2x \frac{dP_n}{dx} + n(n + 1) P_n \right] z^n = 0 \qquad \text{for all } x$$

Therefore

$$\frac{d}{dx} \left[(1 - x^2) \frac{dP_n}{dx} \right] + n(n + 1) P_n = 0$$

Problem If $P_n(x)$ is the nth Legendre polynomial, show that

$$\frac{1}{\sin \theta} \frac{d}{d\theta} \left[\sin \theta \frac{dP_n(\cos \theta)}{d\theta} \right] + n(n + 1) P_n(\cos \theta) = 0$$

Now, if we seek a solution of this differential equation of the form

$$y = \sum_{k=0}^{\infty} C_k x^k$$

we obtain the recursion formula

$$C_{k+2} = \frac{n(n+1) - k(k+1)}{(k+1)(k+2)} \qquad k = 0, 1, 2, \ldots$$

If n is even, the coefficients $C_2, C_4, \ldots, C_n$ will be given by this formula, with $C_{n+2}, C_{n+4}, \ldots$ all vanishing. Similarly, if n is odd, $C_3, C_5, \ldots,$ C_n will be determined, with the remaining odd coefficients vanishing. The resulting terminating power series are the only polynomial solutions of the differential equation. Since the Legendre polynomials $P_n(x)$ are obviously polynomials of degree n, they can be identified with their terminating power series by setting $C_0 = P_n(0)$ in the case of n even and setting $C_1 = dP_n/dx(0)$ in the case of n odd.

Now,

$$P_m(x) \frac{d}{dx}\left[(1-x^2)\frac{dP_n}{dx}\right] - P_n(x)\frac{d}{dx}\left[(1-x^2)\frac{dP_m}{dx}\right]$$
$$+ [n(n+1) - m(m+1)]P_n(x)P_m(x) = 0$$

If we integrate both members of this inequality, we obtain

$$[n(n+1) - m(m+1)] \int_{-1}^{1} P_n(x)P_m(x)dx = 0$$

Hence, because the coefficients of P_n are real,

$$(P_n, P_m) = 0 \qquad \text{if} \qquad n \neq m$$

Returning to (L), we have

$$\frac{1}{1 - 2xz + z^2} = \sum_{n=0}^{\infty} \sum_{m=0}^{\infty} P_n(x)P_m(x)z^{n+m}$$

Taking into account the fact that $(P_n, P_m) = 0$ if $n \neq m$, we integrate both sides of this equality, obtaining

$$\frac{1}{2}\log\frac{1-z}{1+z} = \sum_{n=0}^{\infty}\left\{\int_{-1}^{1}[P_n(x)]^2\,dx\right\}z^{2n}$$

If we now expand the left-hand member of this equality as a power series in z, we obtain

$$\|P_n\|^2 = (P_n, P_n) = \frac{2}{2n+1}$$

Therefore, if $\psi_n = \sqrt{n + \frac{1}{2}}\,P_n$, the ψ_n form an orthonormal sequence.

Problem Show that the functions $\sqrt{n + \frac{1}{2}}\, P_n(\cos\theta)$ form an ortho-normal sequence for the inner product

$$(f,g) = \int_0^\pi f(\theta)g(\theta)\sin\theta\,d\theta$$

Let f be a continuous function of x for $-1 \le x \le 1$. Then, given $\epsilon/\sqrt{2} > 0$, there is a polynomial $p(x)$ such that

$$|f(x) - p(x)| < \frac{\epsilon}{\sqrt{2}}$$

for all x on the interval. This follows from the Weierstrass approximation theorem. Let N be the degree of the polynomial $p(x)$. There then exist constants $a_0, a_1, \ldots, a_n$ such that

$$p(x) = a_0\psi_0 + a_1\psi_1 + \cdots + a_N\psi_N$$

In fact, suppose that there is some linear combination

$$b_0\psi_0 + b_1\psi_1 + \cdots + b_N\psi_N = 0$$

By taking the inner product of the left-hand side with $\psi_0, \psi_1, \ldots, \psi_N$, we obtain, respectively, $b_0 = b_1 = \cdots = b_N = 0$. Therefore the functions $\psi_0, \psi_1, \ldots, \psi_N$ are a linearly independent set of polynomials of degree $<N$, so that every polynomial of degree N can be written as a linear combination of them. But then

$$\left\| f - \sum_{n=0}^N a_n\psi_n \right\| = \|f - p\| = \left[\int_{-1}^1 |f(x) - p(x)|^2\,dx\right]^{\frac{1}{2}} < \epsilon$$

Therefore

$$\left\| f - \sum_{n=0}^N (f,\psi_n)\psi_n \right\| < \epsilon$$

Thus there is a sequence $N_1, N_2, \ldots, N_j, \ldots$ such that

$$\lim_{j\to\infty} \left\| f - \sum_{n=0}^{N_j} (f,\psi_n)\psi_n \right\| = 0$$

or

$$\lim_{j\to\infty} \sum_{n=0}^{N_j} |(f,\psi_n)|^2 = \|f\|^2$$

In this treatment of the Legendre polynomials we obtained many of the properties of the polynomials by considering the differential equation of second order which is satisfied by the polynomials. This is typical of the solutions of a large class of differential equations, the so-called *Sturm-Liouville equations*, which we shall study in subsequent chapters.

Problem

(a) Show that if $f(r,\theta)$ is a function for which

$$\frac{\partial}{\partial r}\left(r^2\frac{\partial f}{\partial r}\right) + \frac{1}{\sin\theta}\frac{\partial}{\partial\theta}\left(\sin\theta\,\frac{\partial f}{\partial\theta}\right) = 0 \qquad 0 \le \theta \le \pi$$

and if

$$a_n(r) = \sqrt{n+\tfrac12}\int_0^\pi f(r,\theta)P_n(\cos\theta)\sin\theta\,d\theta$$

then

$$\frac{d}{dr}\left(r^2\frac{da_n}{dr}\right) - n(n+1)a_n(r) = 0$$

(b) Show that $a_n(r)$ must be of the form

$$C_nr^n + d_nr^{-(n+1)}$$

(c) Show that if $g(\theta),\ 0 \le \theta \le \pi$, is a continuous function, and if

$$g_n = \sqrt{n+\tfrac12}\int_0^\pi g(\theta)P_n(\cos\theta)\sin\theta\,d\theta$$

and if the generalized Fourier series for $f(r,\theta)$ is

$$\sum_{n=0}^{\infty} g_n\frac{r^n}{a^n}\sqrt{n+\tfrac12}\,P_n(\cos\theta)$$

then

$$f(r,\theta) = g(\theta) \qquad \text{for } r = a$$

and $f(r,\theta)$ is bounded for $0 < r \le a$.

(d) Show that if the generalized Fourier series for $f(r,\theta)$ is

$$\sum_{n=0}^{\infty} g_n\frac{a^{n+1}}{r^{n+1}}\sqrt{n+\tfrac12}\,P_n(\cos\theta)$$

then

$$f(r,\theta) = g(\theta) \qquad \text{for } \theta = a$$

and $f(r,\theta)$ is bounded for $r > a$.

(e) Find the generalized Fourier series for $f(r,\theta)$ if

$$f(a,\theta) = g(\theta) \qquad f(b,\theta) = h(\theta) \qquad 0 < a < b$$

Problem   The function $f(r,\theta,\phi)$ is assumed to be a solution of the equation

$$\frac{\partial}{\partial r}\left(r^2\frac{\partial f}{\partial r}\right) + k^2r^2f + \frac{1}{\sin\theta}\frac{\partial}{\partial\theta}\left(\sin\theta\,\frac{\partial f}{\partial\theta}\right) + \frac{1}{\sin^2\theta}\frac{\partial^2 f}{\partial\phi^2} = 0$$

$$0 \le \theta \le \pi,\, 0 \le \phi \le 2\pi$$

We shall not specify the range of the values of r.

(a) If f is assumed to be periodic in ϕ, show that the Fourier coefficient

$$f_m(r,\theta) = \frac{1}{\sqrt{2\pi}} \int_0^{2\pi} f(r,\theta,\phi)e^{-im\phi}\, d\phi$$

satisfies the equation

$$\frac{\partial}{\partial r}\left(r^2 \frac{\partial f_m}{\partial r}\right) + k^2 r^2 f_m + \frac{1}{\sin\theta}\frac{\partial}{\partial\theta}\left(\sin\theta \frac{\partial f_m}{\partial\theta}\right) - \frac{m^2}{\sin^2\theta}f_m = 0$$

(b) If ν is a nonnegative integer, and if $\nu \leq n$, let

$$P_n{}^\nu(x) = (1 - x^2)\frac{\nu}{z}\frac{d^\nu}{dx^\nu}P_n(x)$$

where $P_n(x)$ is the Legendre polynomial. Show that

$$\frac{\partial}{\partial x}\left[(1 - x^2)\frac{d}{dx}P_n{}^\nu(x)\right] + \left[n(n + 1) - \frac{\nu^2}{1 - x^2}\right]P_n{}^\nu(x) = 0$$

Using this differential equation, show that

$$[P_n{}^\nu(x), P_{n'}{}^\nu(x)] = 0 \qquad \text{if } n \neq n'$$

Using the definition of $P_n{}^\nu(x)$, show that

$$[P_n{}^\nu(x), P_n{}^\nu(x)] = [n(n + 1) - \nu(\nu - 1)][P_n{}^{\nu-1}(x), P_n{}^{\nu-1}(x)]$$

and that consequently

$$[P_n{}^\nu(x), P_n{}^\nu(x)] = \frac{2}{2n + 1}\frac{(n + \nu)!}{(n - \nu)!}$$

(c) Show that the functions

$$\psi_n{}^\nu(\theta) = \sqrt{\frac{2n + 1}{2}\frac{(n - \nu)!}{(n + \nu)!}}\, P_n{}^\nu(\cos\theta)$$

form an orthonormal system with respect to the inner product

$$(f,g) = \int_0^\pi f(\theta)\bar{g}(\theta)\sin\theta\, d\theta$$

and that

$$\frac{1}{\sin\theta}\frac{d}{d\theta}\left[\sin\theta\frac{d}{d\theta}\psi_n{}^\nu(\theta)\right] - \frac{m^2}{\sin^2\theta}\psi_n{}^\nu(\theta) = -n(n + 1)\psi_n{}^\nu(\theta)$$

(d) If $f_{m,n}$ is the generalized Fourier coefficient, defined by

$$f_{m,n} = \int_0^\pi f_m(\theta,r)\psi_n{}^{|m|}(\theta)\sin\theta\, d\theta$$

show that

$$\frac{d}{dr}\left(r^2\frac{df_{m,n}}{dr}\right) + [k^2 r^2 - n(n + 1)]f_{m,n}(r) = 0$$

and that if $f_{m,n} = r^{-\frac{1}{2}}\omega_{m,n}$, then

$$r^2 \frac{d^2\omega_{m,n}}{dr^2} + r \frac{d\omega_{m,n}}{dr} + [k^2 r^2 - (n + \frac{1}{2})^2]\omega_{m,n} = 0$$

(e) What is the form of $f_{m,n}$ if $k = 0$?

(f) In addition to what has already been proved, what would be necessary to show that the function f can be represented by the series

$$f = \sum_{n=0}^{\infty} \left[\sum_{m=-n}^{n} \omega_{m,n}(r)\psi_n^{|m|}(\theta)e^{im\phi} \right]$$

(g) If $k = 0$, show that the series given in part (f) can be written as

$$\sum_{n=0}^{\infty} \sum_{m=-n}^{n} \left(\frac{A_{m,n}}{r^{n+1}} + B_{m,n} r^n \right) \psi_n^{|m|}(\theta)e^{im\phi}$$

(h) If $k = 0$, if $f(a,\theta,\phi) = g(\theta,\phi)$, and if $f(r,\theta,\phi) \to 0$ as r becomes large, show that in the region $r > a$, if the series in part (g) converges uniformly to f, then $B_{m,n} = 0$ and

$$A_{m,n} = \frac{a^{n+1}}{2\pi} \int_0^\pi \int_0^{2\pi} g(\theta,\phi)\psi_n^{|m|}(\theta)e^{-im\phi} \, d\phi \, \sin \theta \, d\theta$$

Problem   Let $f(t)$ be given by a Fourier integral,

$$f(t) = \frac{1}{\sqrt{2\pi}} \int_{-\omega}^{\omega} A(\omega)e^{i\omega t} \, d\omega$$

That is, the Fourier transform of $f(t)$ is zero for $|\omega| > W$.

(a) Show that if $A(\omega)$ is piecewise continuous, then

$$A(\omega) = \text{l.i.m.} \sum_{n=-\infty}^{\infty} \sqrt{\frac{\pi}{2}\frac{1}{W}} f\left(\frac{n\pi}{W}\right) e^{-i\frac{n\pi\omega}{W}}$$

(b) Show then that

$$f(t) = \sum_{n=-\infty}^{\infty} f\left(\frac{n\pi}{W}\right) \frac{\sin(Wt - n\pi)}{Wt - n\pi}$$

(c) Show that the functions

$$\psi_n(t) = \sqrt{\frac{W}{\pi}} \frac{\sin(Wt - n\pi)}{Wt - n\pi} \qquad n = 0, \pm1, \pm2, \ldots$$

form an orthonormal sequence with respect to the inner product

$$(f,g) = \int_{-\infty}^{\infty} f(t)\bar{g}(t) \, dt$$

(d) If $g(t)$ is an arbitrary function for which

$$\int_{-\infty}^{\infty} |g(t)|^2 \, dt < \infty$$

what can be said about the relationship between $g(t)$ and its generalized Fourier series with respect to the orthonormal sequence $\{\psi_n(t)\}$?

8
The Sturm-Liouville Problem

Let us consider the differential equation

$$-\frac{d^2y}{dx^2} + q(x)y = \lambda y \qquad -\infty < a \le x \le b < \infty \tag{S-L}$$

The function q is assumed to be real and continuous on the closed interval. We seek solutions to (S-L) which satisfy one of two sets of conditions:

$$\begin{aligned} \sin \alpha \, y(a) - \cos \alpha \, y'(a) &= 0 \qquad 0 \le \alpha \le \pi \\ \sin \beta \, y(b) - \cos \beta \, y'(b) &= 0 \qquad 0 \le \beta \le \pi \end{aligned} \tag{B}$$

and

$$y(a) = y(b) \qquad y'(a) = y'(b) \tag{P}$$

In general, nonzero solutions of (S-L) which satisfy (B) or (P) exist only for certain special values of λ, which are called *characteristic values*. If λ is a characteristic value and y is a solution, y is called a *characteristic function.* The problem of finding the characteristic functions and characteristic values is called a *regular Sturm-Liouville problem.* If conditions (B) are imposed, we are dealing with the *homogeneous boundary-value problem.* If conditions (P) are imposed, we are dealing with the *periodic problem.* There are other regular Sturm-Liouville problems, called the *general self-adjoint problem,* of which (B) and (P) are special cases.

If $q(x) = 0$, and we solve (P) with $a = 0$ and $b = L$, we find that the general solution of (S-L) is of the form

$$y = A \sin \sqrt{\lambda} x + B \cos \sqrt{\lambda} x$$

In order that (P) be satisfied, it is necessary and sufficient that

$$B(\cos \sqrt{\lambda}\, L - 1) + A \sin \sqrt{\lambda}\, L = 0$$
$$-B \sqrt{\lambda} \sin \sqrt{\lambda}\, L + A \sqrt{\lambda}\, (\cos \sqrt{\lambda}\, L - 1) = 0$$

This system will have a solution for which not both A and B vanish if and only if

$$\sqrt{\lambda}\, (1 - \cos \sqrt{\lambda}\, L) = 0$$

The only solutions to this equation are

$$\lambda_n = \frac{4n^2\pi^2}{L^2} \qquad n = 0, 1, 2, \ldots$$

For such values of λ_n, A and B are arbitrary. Thus for each characteristic value λ_n there are two linearly independent characteristic functions, $\sin (2\pi nx/L)$ and $\cos (2\pi nx/L)$, or, equivalently,

$$\frac{e^{\pm 2\pi nix/L}}{\sqrt{L}}$$

We have obtained a representation of the characteristic functions which is identical with the functions studied in connection with Fourier series. In general, the characteristic functions for a regular Sturm-Liouville problem form a sequence $\{\psi_n\}$ that is orthonormal, for

$$(f,g) = \int_a^b f\bar{g}\, dx$$

and is complete in the class of functions which satisfy (B) or (P), depending upon which problem we are dealing with. These facts we already know to be true in the case just described. It is their general establishment with which we are concerned in this chapter.

Problem   Consider the partial-differential equation

$$\frac{\partial u}{\partial t} = k^2 \frac{\partial^2 u}{\partial x^2}$$

Suppose a solution of this equation, $u(x,t)$, is sought for which

$$u(0,t) = u(L,t) = 0$$

and which is of the form

$$u(x,t) = T(t)X(x)$$

(a) Show that

$$\frac{T'(t)}{T(t)} = k^2 \frac{X''(x)}{X(x)}$$

(b) Show now that

$$-X''(x) = \lambda X$$

where λ is a constant, and that

$$X(0) = X(L) = 0$$

(c) Show that the only values of λ for which the conditions of part (b) are satisfied are given by

$$\lambda_n = \frac{n^2\pi^2}{L^2} \qquad n = 1, 2, \ldots$$

and that consequently the only solutions $u(x,t)$ satisfying the given conditions are of the form

$$u_n(x,t) = C_n e^{-k^2 n^2 \pi^2 t/L^2} \sin \frac{n\pi x}{L}$$

where C_n is a constant.

(d) What are the solutions if

$$\frac{\partial u(0,t)}{\partial x} = \frac{\partial u(L,t)}{\partial x} = 0$$

(e) What are the solutions if

$$u(0,t) = 0 \qquad \text{and} \qquad \frac{\partial u(L,t)}{\partial x} = 0$$

Problem Find the most general form of the solution of the equation

$$C^2 \frac{\partial^2 u}{\partial x^2} - \frac{\partial^2 u}{\partial t^2} = 0$$

which is of the form

$$u(x,t) = T(t)X(x)$$

and for which

$$u(0,t) = u(L,t) = 0$$

Problem For what values of k does the equation

$$\frac{\partial^2 u}{\partial x^2} + \frac{\partial^2 u}{\partial y^2} + k^2 u = 0$$

have solutions of the form

$$u(x,y) = X(x)X(y)$$

where

$$u\left(-\frac{a}{2}, y\right) = u\left(\frac{a}{2}, y\right) = 0$$

$$\frac{\partial u(x, b/2)}{\partial y} = \frac{\partial u(x, -b/2)}{\partial y} = 0$$

What is the general form of the solutions?

Answer

$$k^2 = \left(\frac{m^2}{b^2} + \frac{n^2}{a^2}\right)\pi^2 \qquad \begin{array}{l} n = 1, 2, 3, \ldots \\ m = 0, 1, 2, \ldots \end{array}$$

$$u(x,y) = C_{n,m} \sin \frac{n\pi(a - 2x)}{2a} \cos \frac{m\pi(b - 2y)}{2b}$$

where $C_{n,m}$ is a constant.

Problem Find the most general form of the solutions of

$$r^2 \frac{\partial^2 u}{\partial r^2} + r \frac{\partial u}{\partial r} + \frac{\partial^2 u}{\partial \theta^2} = 0$$

which are in the form

$$u(r,\theta) = f(r)g(\theta)$$

and for which

$$u(r,\alpha) = u(r,-\alpha) = 0$$

Answer

$$u(r,\theta) = \sin \frac{n\pi(\alpha - \theta)}{2\alpha} (a_n r^{n\pi/2\alpha} + b_n r^{-n\pi/2\alpha})$$

where a_n and b_n are constants.

Let y_1 and y_2 satisfy (B). Then it is clear that every linearly independent combination of y_1 and y_2 also satisfies (B). Therefore if y_1

and y_2 are characteristic functions for the characteristic value λ, then y_1 and y_2 must be linearly dependent; otherwise *every* solution of (S-L) would satisfy (B). There does exist a solution $y(x)$ for which $y(a) = \sin \alpha$ and $y'(a) = -(\cos \alpha)$; in fact, the values of y and y' at $x = a$ can be chosen arbitrarily. But this solution does not satisfy (B). Therefore, *for the homogeneous boundary-value problem, except for a multiplicative constant, for each characteristic value there is exactly one characteristic function.* We have already seen that for the periodic problem this is not necessarily the case.

RESTATEMENT OF THE STURM–LIOUVILLE PROBLEM

We shall show that all the characteristic values are real, and that there exists a real number Λ such that if λ is a characteristic value, then $\lambda > \Lambda$. First, let us assume that λ is a characteristic value and that y is a characteristic function for λ. Then, multiplying (S-L) by $\bar{y}$ and integrating from a to b, we have

$$\lambda \int_a^b |y|^2 \, dx = \int_a^b q(x)|y|^2 \, dx + \int_a^b |y'(x)|^2 \, dx - \left[\bar{y}(x) \frac{dy(x)}{dx} \right]_a^b$$

For the periodic problem, conditions (P), the last term on the right-hand side of this equation vanishes, so that clearly λ is real, and $\lambda > \min q(x)$. For problem (B), there is a constant C_a such that

$$y(a) = C_a \cos \alpha \qquad y'(a) = C_a \sin \alpha$$

and a constant C_b such that

$$y(b) = C_b \cos \beta \qquad y'(b) = C_b \sin \beta$$

Then the last term on the right-hand side of this equation is

$$|C_a|^2 \sin \alpha \cos \alpha - |C_b|^2 \sin \beta \cos \beta$$

which is real. Consequently, λ is real.

If $\sin \alpha \cos \alpha = \sin \beta \cos \beta = 0$, it is still true, as it was for (P), that $\lambda > \min q(x)$. In order to complete the argument, we must resort to a less elementary analysis. We shall assume that $\lambda = -p^2$, $p > 0$, and we shall establish that p cannot be arbitrarily large. Let $C_a = 1$. Then y satisfies the equation

$$y(x) = \cos \alpha \cos p(x - a) + \frac{1}{p} \sin \alpha \sin p(x - a)$$

$$+ \frac{1}{p} \int_a^x \sin p(x - \zeta) q(\zeta) y(\zeta) \, d\zeta$$

Let

$$v(x) = e^{-p(x-a)}y(x)$$

$$= \frac{1}{2}\left[\cos\alpha\,(1 + e^{-2p(x-a)}) + \frac{1}{p}\sin\alpha\,(1 - e^{-2p(x-a)})\right.$$

$$\left. + \frac{1}{p}\int_a^x (1 - e^{-2p(x-\zeta)})q(\zeta)v(\zeta)\,d\zeta\right]$$

Then

$$|v(x)| \le B(\alpha,p) + \frac{1}{2p}\int_a^x |q(\zeta)|\,|v(\zeta)|\,d\zeta = H(x)$$

where

$$B(\alpha,p) = \sqrt{\cos^2\alpha + \frac{1}{4p^2}\sin^2\alpha}$$

Now

$$H(a) = B(\alpha,p)$$

$$\frac{dH}{dx} = \frac{1}{2p}|q(x)|\,|v(x)| \le \frac{1}{2p}|q(x)|H(x)$$

and so

$$H(x) \le B(\alpha,p)\exp\left[\frac{1}{2p}\int_a^x |q(\zeta)|\,d\zeta\right]$$

Therefore, if $A > 1$, there exists a $p_0 > 0$ such that if $p > p_0$, then

$$|v(x)| < AB(\alpha,p)$$

We may then write

$$y(b) = \frac{1}{2}e^{p(b-a)}\left[\cos\alpha\,(1 + e^{-2p(b-a)}) + \frac{C_1(p)}{p}\right]$$

$$y'(b) = \frac{1}{2}pe^{p(b-a)}\left[\cos\alpha\,(1 - e^{-2p(b-a)}) + \frac{C_2(p)}{p}\right]$$

where

$$C_1(p) = \sin\alpha\,(1 - e^{-2p(b-a)}) + \int_a^b (1 - e^{-2p(b-\zeta)})q(\zeta)v(\zeta)\,d\zeta$$

$$C_2(p) = \sin\alpha\,(1 + e^{-2p(b-a)}) + \int_a^b (1 + e^{-2p(b-\zeta)})q(\zeta)v(\zeta)\,d\zeta$$

Since $\lambda = -p^2$ is a characteristic value,

$$\sin\beta\,y(b) - \cos\beta\,y'(b) = 0$$

or

$$p\cos\beta\left[\cos\alpha\,(1 - e^{-2p(b-a)}) + \frac{C_2(p)}{p}\right]$$

$$= \sin\beta\left[\cos\alpha\,(1 + e^{-2p(b-a)}) + \frac{C_1(p)}{p}\right]$$

We shall now show that this equation cannot be satisfied by arbitrarily large values of p.   Let us first consider the situation for $\cos \alpha \neq 0$.   If $\cos \beta \neq 0$, then

$$1 - e^{-2p(b-a)} = \frac{\tan \beta}{p \cos \alpha} \left[\cos \alpha \, (1 + e^{-2p(b-a)}) + \frac{C_1(p)}{p} \right] - \frac{C_2(p)}{p \cos \alpha}$$

As $p \to \infty$, the right-hand side of this equation approaches zero and the left-hand side approaches 1; for when $p > p_0$,

$$|C_1(p)| \leq |\sin \alpha| + AB(\alpha,p) \int_a^b |q(\zeta)| \, d\zeta$$

$$|C_2(p)| \leq 2 \left[|\sin \alpha| + AB(\alpha,p) \int_a^b |q(\zeta)| \, d\zeta \right]$$

and therefore $C_1(p)$ and $C_2(p)$ remain bounded as $p \to \infty$.   Thus there must be a value of p, $p_1 > 0$, such that if $p > p_1$, then $\lambda = -p^2$ is not a characteristic value.   If $\cos \beta = 0$, the required condition is

$$1 + e^{-2p(b-a)} = - \frac{C_1(p)}{\cos \alpha \, p}$$

and similar remarks apply.

If $\cos \alpha = 0$, the condition becomes

$$\cos \beta \, C_2(p) = \sin \beta \, \frac{C_1(p)}{p}$$

or

$$\cos \beta \sin \alpha \, (1 + e^{-2p(b-a)}) = \sin \beta \, \frac{C_1(p)}{p}$$

$$- \cos \beta \int_a^b (1 + e^{-2b(b-\zeta)}) q(\zeta) v(\zeta) \, d\zeta$$

because when $\cos \alpha = 0$ and $|B(\alpha,p)| \leq 1/2p$, then the right-hand side of this equation approaches zero as $p \to \infty$, and the left-hand side approaches $\cos \beta$.

If $\cos \beta = 0$, we already know that $\lambda > \min q(x)$.   If $\cos \beta \neq 0$, values of p and p_2 must exist such that if $p > p_2$, then $\lambda = -p^2$ is not a characteristic value.

Now, in the differential equation (S-L), if we replace $q(x)$ by $q(x) - \Lambda$, the characteristic value λ is replaced by $\lambda - \Lambda$.   Thus we have transformed the original Sturm-Liouville problem to an equivalent one in which all the characteristic values are positive.   It will greatly simplify the rest of our discussions if we assume that all the characteristic values of the Sturm-Liouville problem are positive.   No generality is lost by this assumption.

Problem   Let $u(x)$ be a function which has two continuous derivatives and which satisfies the conditions (B) or (P).   Let $v(x)$ be a function which has one continuous derivative and for which

$$v(a) = v'(a) = v(b) = v'(b) = 0$$

For any function ω with continuous first derivatives, let

$$P(\omega) = \frac{\displaystyle\int_a^b \{(d\omega/dx)^2 + q(x)[\omega(x)]^2\}\,dx}{\displaystyle\int_a^b [\omega(x)]^2\,dx}$$

(a) Show that

$$\frac{d}{d\epsilon}P(u + \epsilon v)\bigg]_{\epsilon=0} = 0$$

if and only if

$$\int_a^b \left\{-\frac{d^2u}{dx^2} + [q(x) - \lambda]u\right\}v = 0$$

where $\lambda = P(u)$.

(b) Show that the smallest characteristic value λ_0 can be obtained by finding a function u_0 which satisfies the boundary conditions and for which $P(u_0) \leq P(u)$, where u is any other function satisfying the boundary conditions, and show that $\lambda_0 = P(u_0)$.

THE INTEGRAL EQUATION

Let us suppose that λ is a characteristic value for the homogeneous boundary-value problem and y is the associated characteristic function. Let $y_1(x)$ and $y_2(x)$ be two solutions of the equation

$$-\frac{d^2y}{dx^2} + q(x)y = 0$$

which satisfy the conditions

$$\begin{array}{ll} y_1(a) = \cos\alpha & y_1'(a) = \sin\alpha \\ y_2(b) = \cos\beta & y_2'(b) = \sin\beta \end{array}$$

Then y_1 satisfies the first condition in (B) and y_2 satisfies the second. The functions y_1 and y_2 must be linearly independent, since otherwise y_1 would be either identically zero or a constant multiple of y_2. The first of these is impossible because not both $y_1(a)$ and $y_1'(a)$ are zero. The second is impossible because then y_1 would satisfy both requirements of (B), which would mean that $\lambda = 0$ was a characteristic value. Therefore

$$W(y_1, y_2) = y_1 y_2' - y_2 y_1'$$

is a nonzero constant.

Let f be continuous for $a \leq x \leq b$, and let g be defined as

$$g(x) = -\frac{1}{W(y_1, y_2)} \left[\int_a^x y_2(x)y_1(\zeta)f(\zeta) \, d\zeta + \int_x^b y_1(x)y_2(\zeta)f(\zeta) \, d\zeta \right]$$

It is easy to verify that

$$-\frac{d^2g}{d\lambda^2} + q(x)g = f(x)$$

$$g(a) = -\left[\frac{1}{W(y_1, y_2)} \int_a^b y_2(\zeta)f(\zeta) \, d\zeta \right] y_1(a)$$

$$g'(a) = -\left[\frac{1}{W(y_1, y_2)} \int_a^b y_2(\zeta)f(\zeta) \, d\zeta \right] y_1'(a)$$

$$g(b) = -\left[\frac{1}{W(y_1, y_2)} \int_a^b y_1(\zeta)f(\zeta) \, d\zeta \right] y_2(b)$$

$$g'(b) = -\left[\frac{1}{W(y_1, y_2)} \int_a^b y_1(\zeta)f(\zeta) \, d\zeta \right] y_2'(b)$$

These last four equalities show that g satisfies conditions (B). Let us rewrite g as

$$g(\zeta) = \int_a^b G(x,\zeta)f(\zeta) \, d\zeta$$

where

$$G(x,\zeta) = \begin{cases} -\dfrac{1}{W(y_1, y_2)} f_1(\zeta)f_2(x) & a \leq \zeta \leq x \leq b \\[2mm] -\dfrac{1}{W(y_1, y_2)} f_2(\zeta)f_1(x) & a \leq x \leq \zeta \leq b \end{cases}$$

If y is a continuous function for which

$$y(x) = \lambda \int_a^b G(x,\zeta)y(\zeta) \, d\zeta \tag{I}$$

then $y(x)$ is twice differentiable and satisfies equation (S-L) and conditions (B). Conversely, if y satisfies equation (S-L) and conditions (B), then we have

$$\lambda \int_a^b y(\zeta)G(x,\lambda) \, d\zeta = \int_a^b \left[-\frac{d^2y}{d\zeta^2} + q(\zeta)y(\zeta) \right] G(\zeta,x) \, d\zeta$$

It is an exercise in integration by parts to show that the right-hand member of this equality is $y(x)$. Therefore the characteristic values and the characteristic functions for the homogeneous boundary-value problem are precisely those values of λ and functions y which satisfy equation (I), which is called an *integral equation*.

We shall investigate some of the properties of this integral equation. If y is a continuous function on $a \leq x \leq b$, then we define the function

Ty by the formula

$$Ty = \int_a^b G(x,\zeta)y(\zeta)\,d\zeta$$

Theorem 1 Let Ty be as defined above; then Ty is a continuous function on $a \le x \le b$. If y_1 and y_2 are continuous functions on the interval, and if c_1 and c_2 are constants, then

$$T(c_1y_1 + c_2y_2) = c_1Ty_1 + c_2Ty_2$$

If

$$(f,g) = \int_a^b f\bar{g}\,dx, \text{ then for } y \text{ and } z \text{ continuous}$$
$$(Ty,z) = (y,Tz)$$

There exists a constant M such that for all continuous y,

$$\|Ty\| \le M\|y\|$$

If $y_1, y_2, \ldots, y_n, \ldots$ is a sequence of continuous functions for which $\|y_n\| \le A$ for all n, then $\{z_n\}$, $z_n = Ty_n$, is a uniformly equicontinuous, uniformly bounded sequence; that is, there is a constant C such that for all x in $a \le x \le b$,

$$|z_n(x)| < C \qquad \text{for all } n$$

and for every $\epsilon > 0$ there exists a $\delta > 0$ such that if x_1 and x_2 lie in the interval and $|x_1 - x_2| < \delta$, then

$$|z_n(x_1) - z_n(x_2)| < \epsilon \qquad \text{for all } n$$

Proof The fact that $T(c_1y_1 + c_2y_2) = c_1Ty_1 + c_2Ty_2$ follows directly from the representation of Ty as an integral. By writing (Ty,z) as an iterated integral, reversing the order of integration, interchanging the variables of integration x and ζ, and making use of the fact that $G(x,\zeta)$ is real and $G(x,\zeta) = G(\zeta,x)$, which is apparent from the definition of $G(x,\zeta)$, we obtain

$$(Ty,z) = (y,Tz)$$

From the Cauchy-Schwartz inequality,

$$|Ty(x)| \le \left\{\int_a^b [G(x,\zeta)]^2\,d\zeta\right\}^{\frac{1}{2}} \|y\|$$

and so

$$\|Ty\| \le \left\{\iint_a^b [G(x,\zeta)]^2\,d\zeta\,dx\right\}^{\frac{1}{2}} \|y\|$$

Therefore, taking

$$M = \left\{ \int_a^b \!\!\int [G(x,\zeta)]^2 \, d\zeta \, dx \right\}^{\frac{1}{2}}$$

which exists since $G(x,\zeta)$ is continuous on the rectangle $a \le x \le b$, $a \le \zeta \le b$, we have

$$\|Ty\| \le M\|y\|$$

If we let $z(x) = Ty(x)$, where $\|y\| < A$, again using the Cauchy-Schwartz inequality, we have

$$|z(x_1) - z(x_2)| \le \left\{ \int_a^b [G(x_1,\zeta) - G(x_2,\zeta)]^2 \, d\zeta \right\}^{\frac{1}{2}} A$$

Since $G(x,\zeta)$ is continuous on the rectangle, it is uniformly continuous on the rectangle. Given $\epsilon > 0$, there exists a $\delta > 0$ such that for x_1 and x_2 in the interval $a \le x \le b$ and $a \le \zeta \le b$, if $|x_1 - x_2| < \epsilon$,

$$|G(x_1,\zeta) - G(x_2,\zeta)| < \frac{\epsilon}{A(b-a)^{\frac{1}{2}}}$$

Then

$$|z(x_1) - z(x_2)| < \epsilon$$

It is also obvious that

$$|z(x)| \le C$$

where

$$C = \max_{a \le x \le b} \left\{ \int_a^b G(x,\zeta)]^2 \, d\zeta \right\}^{\frac{1}{2}} A$$

Problem

(a) Find $G(x,\xi)$ such that if

$$g(x) = \int_0^L G(x,\xi)f(\xi) \, d\xi$$

then

$$-\frac{d^2g}{dx^2} = f(x) \qquad g(0) = 0 \qquad \frac{dg}{dx}(L) = 0$$

(b) Show that the characteristic functions of this Sturm-Liouville problem are

$$y_n(x) = \sqrt{\frac{2}{L}} \sin \lambda_n x \qquad \lambda_n = \frac{(n + \frac{1}{2})\pi}{L}$$

and that the generalized Fourier series for $G(x,\xi)$, in terms of the orthonormal sequence $\{y_n\}$, is

$$G(x,\xi) \sim \sum_{n=1}^{\infty} \frac{y_n(x)y_n(\xi)}{\lambda_n}$$

(c) Show that

$$\left[\int_0^L |g(x)|^2 \, dx \right]^{1/2} \leq \frac{L^2}{\sqrt{6}} \left[\int_0^L |f(x)|^2 \, dx \right]^{1/2}$$

For the periodic problem, conditions (P), a similar result holds. The problem is equivalent to an integral equation of the form (I), and the function Ty has the same properties as in the above theorem. All that is involved is finding the function $G(x,\zeta)$ and showing that it is real and continuous on the rectangle and that $G(x,\zeta) = G(\zeta,x)$.

In order to do this, let y_1 and y_2 be the solutions of

$$-\frac{d^2y}{dx^2} + q(x)y = 0$$

for which

$$y_1(a) = 1 \qquad y_1'(a) = 0$$
$$y_2(a) = 0 \qquad y_2'(a) = 1$$

Then y_1 and y_2 are linearly independent and real, and their wronskian $W(y_1,y_2)$ is identically equal to 1.   The general solution of the equation

$$-\frac{d^2g}{dx^2} + q(x)g = f$$

is of the form

$$g(x) = \alpha y_1(x) + \beta y_2(x) - \int_a^x [y_2(x)y_1(\zeta) - y_1(x)y_2(\zeta)]f(\zeta) \, d\zeta$$

The function $g(x)$ will satisfy conditions (P) if and only if α and β satisfy the equations

$$[y_1(b) - 1]\alpha + y_2(b)\beta = \int_a^b [y_2(b)y_1(\zeta) - y_1(b)y_2(\zeta)]f(\zeta) \, d\zeta$$
$$y_1'(b)\alpha + [y_2'(b) - 1]\beta = \int_a^b [y_2'(b)y_1(\zeta) - y_1'(b)y_2(\zeta)]f(\zeta) \, d\zeta$$

The determinant

$$\Delta = 2 - y_1(b) - y_2'(b)$$

of this system is not zero; if it were, there would exist an α and a β, not both zero, such that $\alpha y_1 + \beta y_2$ satisfied (P), which would mean that

$\lambda = 0$ was a characteristic value. Thus an α and a β can be found which cause g to satisfy (P). Then g can be written as

$$g(x) = \int_a^b G(x,\zeta)f(\zeta) \, d\zeta$$

where

$$G(x,\zeta) = -\frac{1}{\Delta} \{y_2(b)y_1(\zeta)y_1(x) + [1 - y_1(b)]y_1(\zeta)y_2(x)$$
$$- [1 - y_2'(b)]y_2(\zeta)y_1(x) - y_1'(b)y_2(\zeta)y_2(x)\} \qquad \text{for } a \leq \zeta \leq x \leq b$$

and

$$G(\zeta,x) = G(x,\zeta)$$

CHARACTERISTIC VALUES AND FUNCTIONS

We have now established that for both the homogeneous boundary-value problem and the periodic problem the number λ will be a characteristic value and the continuous function y will be a characteristic function if and only if

$$Ty = \mu y$$

where Ty has the properties enunciated above and $\mu = 1/\lambda$. In order to capitalize on these properties, we shall first prove *Azela's theorem*.

Theorem 2 Let $\{z_n\}$ be a sequence of functions which is equicontinuous and bounded at each point of an interval $a \leq x \leq b$. Then there exists a subsequence $\{z_{n_j}\}$ which converges uniformly on $a \leq x \leq b$.

Proof Let $\{x_k\}$ be an enumeration of the rational numbers in $a \leq x \leq b$. We shall define, for each integer k, a sequence of integers $\{\pi(k;j)\}$. For $k = 1$ we first examine the sequence of numbers

$$z_1(x_1), z_2(x_1), \ldots, z_n(x_1), \ldots$$

Since this is a bounded sequence, it has a convergent subsequence

$$z_{n_1}(x_1), z_{n_2}(x_1), \ldots, z_{n_j}(x_1), \ldots$$

Let $\pi(1;j)$ be defined to be n_j. Now, suppose that for a fixed value of k the sequence $\pi(k,j)$ has been defined so that the sequence

$$z_{\pi(k,1)}(x_k), \ldots, z_{\pi(k,n)}(x_k), \ldots$$

converges. Consider the sequence

$$z_{\pi(k,1)}(x_{k+1}), \ldots, z_{\pi(k,n)}(x_{k+1}), \ldots$$

Since this is a bounded sequence, it has a convergent subsequence

$$z_{\pi(k,n_1)}(x_{k+1}), \; \ldots \; , z_{\pi(k,n_j)}(x_{k+1}), \; \ldots$$

Now define $\pi(k + 1, j)$ to be $\pi(k,n_j)$. The numbers $\pi(k,j)$ are thus defined, by induction, for all k and all j. We see that the sequence $\{\pi(k + 1, j)\}$ is a subsequence of $\{\pi(k,j)\}$, and that the sequence $\{z_{\pi(k,j)}(x_k)\}$ converges. Now consider the sequence $\{n_j\}$, defined by the formula

$$n_j = \pi(j,j)$$

For fixed k_1, the sequence $\{n_j\}$ for $j \geq k$ is a subsequence of the sequence $\{\pi(k,j)\}$. Therefore, since the sequence $\{z_{n_j}(x_k)\}$ is a subsequence of a convergent sequence, it is convergent.

We have thus exhibited a subsequence $\{z_{n_j}\}$ of our original sequence of functions which converges at each of the points x_k. It remains to demonstrate that this same subsequence actually converges uniformly on the interval $a \leq x \leq b$.

Let x be a point in the interval. We wish to show that for $\epsilon > 0$ there exists a J such that if $j_1 > J$ and $j_2 > J$, then

$$|z_{n_j}(x) - z_{n_{j_2}}(x)| < \epsilon$$

From the equicontinuity, there is a $\delta > 0$ such that if $|x' - x| < \delta$, then

$$|z_n(x') - z_n(x)| < \frac{\epsilon}{3} \qquad \text{for all } n$$

Let x_k be any rational number such that $|x_k = x| < \delta$. Let us keep x_k fixed. Because the sequence $\{z_{n_j}(x_k)\}$ converges, there is a J such that if $j_1 > J$ and $j_2 > J$, then

$$|z_{n_{j_1}}(x_k) - z_{n_{j_2}}(x_k)| < \frac{\epsilon}{3}$$

But then, if $j_1 > J$ and $j_2 > J$,

$$|z_{n_{j_1}}(x) - z_{n_{j_2}}(x)| \leq |z_{n_{j_1}}(x) - z_{n_{j_1}}(x_k)| + |z_{n_{j_1}}(x_k) - z_{n_{j_2}}(x_k)|$$
$$+ |z_{n_{j_2}}(x_k) - z_{n_{j_2}}(x)| < \frac{\epsilon}{3} + \frac{\epsilon}{3} + \frac{\epsilon}{3} = \epsilon$$

Therefore the sequence $\{z_{n_j}\}$ converges at each x in the interval.

To show uniform convergence, we must prove that for $\epsilon > 0$ there exists a J such that if $j_1 > J$ and $j_2 > J$, then

$$|z_{n_{j_1}}(x) - z_{n_{j_2}}(x)| < \epsilon \qquad \text{for all } x$$

We prove this by the method of contradiction. Suppose that the sequence is not uniformly convergent. Then there would exist an $\epsilon > 0$ such that for every integer J there would exist two integers, $j_1(J)$ and $j_2(J)$, both greater than J, and a point x_J such that

$$|z_{n_{j_1(J)}}(x_J) - z_{n_{j_2(J)}}(x_J)| \geq \epsilon$$

The sequence of points $\{x_J\}$ has a subsequence $\{x_{J_k}\}$ which converges to some point x^* in the interval. There is an integer J' such that if $j_1 > J'$ and $j_2 > J'$,

$$|z_{n_{j_1}}(x^*) - z_{n_{j_2}}(x^*)| < \frac{\epsilon}{3}$$

There is also a $\delta > 0$ such that if $|x' - x^*| < \delta$,

$$|z_n(x') - z_n(x^*)| < \frac{\epsilon}{3} \qquad \text{for all } n$$

Now let k be such that

$$J_k > J' \qquad \text{and} \qquad |x_{J_k} - x^*| < \delta$$

Then

$$|z_{n_{j_1(J_k)}}(x_{J_k}) - z_{n_{j_2(J_k)}}(x_{J_k})| \leq |z_{n_{j_1(J_k)}}(x_{J_k}) - z_{n_{j_1(J_k)}}(x^*)|$$
$$+ |z_{n_{j_1(J_k)}}(x^*) - z_{n_{j_2(J)}}(x^*)| + |z_{n_{j_2(J_k)}}(x^*) - z_{n_{j_2(J_k)}}(x_{J_k})|$$
$$< \frac{\epsilon}{3} + \frac{\epsilon}{3} + \frac{\epsilon}{3} = \epsilon$$

which is a contradiction.

We now can produce the characteristic values and the characteristic functions. We know that for $\|y\| = 1$ there is a constant M such that $\|Ty\| \leq M$. We define $\|T\|$ as the least upper bound of the set of real numbers $\|Ty\|$, in which $\|y\| = 1$. That is, for $\|y\| = 1$, $\|Ty\| \leq \|T\|$, and given $\epsilon > 0$, there exists a y, $\|y\| = 1$, such that $\|T\| < \|Ty\| + \epsilon$.

Theorem 3 There exists a continuous function y for which

$$\|y\| = 1 \qquad \text{and} \qquad \|Ty\| = \|T\| \, \|y\|$$

Proof For any continuous function y

$$|(Ty, y)| \leq \|Ty\| \, \|y\| \leq \|T\| \, \|y\|^2$$

Therefore, if $\|y\| = 1$,

$$|(Ty, y)| \leq \|T\|$$

Let M be the greatest lower bound of the set of numbers $|(Ty,y)|$, $\|y\| = 1$. We have just observed that $M \leq \|T\|$. We shall now show that $M = \|T\|$. For any continuous y it is easy to verify that for $c \neq 0$,

$$\|Ty\|^2 = \frac{1}{4}\left[\left(T\left(cy + \frac{1}{c}Ty\right), cy + \frac{1}{c}Ty\right)\right.$$
$$\left. - \left(T\left(cy - \frac{1}{c}Ty\right), cy - \frac{1}{c}Ty\right)\right]$$

Then

$$\|Ty\|^2 \leq \frac{1}{4}M\left(\left\|cy + \frac{1}{c}Ty\right\|^2 + \left\|cy - \frac{1}{c}Ty\right\|^2\right)$$
$$= \frac{M}{2}\left(c^2\|y\|^2 + \frac{1}{c^2}\|Ty\|^2\right)$$

Now, for $\epsilon > 0$, there exists a y, $\|y\| = 1$, such that $\|Ty\| > \|T\| - \epsilon$. Let $c^2 = \|Ty\|$. Then, from the above inequality,

$$M \geq \|T\| - \epsilon \qquad \text{for all } \epsilon$$

and so $M \geq \|T\|$. There then exists a sequence of continuous functions $\{y_n\}$ for which $\lim_{n \to \infty} |(Ty_n, y_n)| = \|T\|$

Therefore a subsequence of $\{(Ty_n, y_n)\}$ converges to a number μ_0, $|\mu_0| = \|T\|$. Without loss of generality, we shall assume that this subsequence is $\{(Ty_n, y_n)\}$. If $z_n = Ty_n$, the sequence $\{z_n\}$ is equicontinuous and bounded at each point of $a \leq x \leq b$. Hence a subsequence converges uniformly on $a \leq x \leq b$. Again without loss of generality, we assume that the subsequence is $\{z_n\}$. Let the limit function be z. We observe that

$$\|Ty_n - \mu_0 y_n\|^2 = (Ty_n, Ty_n) - 2\mu_0(Ty_n, y_n) + \mu_0^2$$
$$\leq 2\|T\|^2 - 2\mu_0(Ty_n, y_n)$$

Therefore

$$\lim_{n \to \infty} \|Ty_n - \mu_0 y_n\| = 0$$

But

$$Tz_n - \mu_0 z_n = T(Ty_n - \mu_0 y_n)$$

and

$$\|Tz_n - \mu_0 z_n\| \leq \|T\| \, \|Ty_n - \mu_0 y_n\|$$

Thus

$$\lim_{n \to \infty} \|Tz_n - \mu_0 z_n\| = 0$$

We shall now show that

$$Tz = \mu_0 z$$

We first prove an elementary inequality called the *triangle inequality*. If y_1 and y_2 are any two continuous functions, then

$$\|y_1 + y_2\|^2 = \|y_1\|^2 + 2(y_1, y_2) + \|y_2\|^2 \leq \|y_1\|^2 + 2\|y_1\| \|y_2\| + \|y_2\|^2$$
$$= (\|y_1\| + \|y_2\|)^2$$

Therefore

$$\|y_1 + y_2\| \leq \|y_1\| + \|y_2\|$$

By an elementary induction,

$$\|y_1 + y_2 + \cdots + y_n\| \leq \|y_1\| + \|y_1\| + \cdots + \|y_n\|$$

Then

$$\|Tz - \mu_0 z\| = \|(Tz - Tz_n) + (Tz_n - \mu_0 z_n) + (\mu_0 z_n - \mu_0 z)\|$$
$$\leq \|Tz - Tz_n\| + \|Tz_n - \mu_0 z_n\| + \|\mu_0 z_n - \mu_0 z\|$$
$$\leq (\|T\| + |\mu_0|)\|z_n - z\| + \|Tz_n - \mu_0 z_n\|$$

We have already proved that

$$\lim_{n \to \infty} \|Tz_n - \mu_0 z_n\| = 0$$

Since the sequence $\{z_n\}$ converges uniformly to z, the sequence of functions $|z_n - z|^2$ converges uniformly to zero, and so

$$\lim_{n \to \infty} z_n - z = 0$$

Therefore $\|Tz - \mu_0 z\| = 0$. But if

$$\int_a^b |Tz - \mu_0 z|^2 \, dx = 0$$

the function $|Tz - \mu_0 z|^2$, being continuous and nonnegative, must be identically zero. Therefore $Tz = \mu_0 z$. If $\|z\| = 0$, then

$$\lim \|Ty_n\| = 0$$

But

$$\mu_0 y_n = (\mu_0 y_n - Ty_n) + Ty_n$$

and so

$$\lim \|\mu_0 y_n\| = |\mu_0| = 0$$

This would imply that $\|T\| = 0$, or $Ty = 0$, for every y. Thus $\|z\| \neq 0$, so that if $y = z/\|z\|$, we have

$$\|y\| = 1 \qquad \text{and} \qquad Ty = \mu_0 y$$

where $|\mu_0| = \|T\|$. Therefore $\lambda_0 = 1/\mu_0$ is a characteristic value, and since it is positive, we have $\mu_0 = \|T\|$.

We have just proved the existence of a function y_0 such that

$$\|y_0\| = 1 \qquad \text{and} \qquad Ty_0 = \mu_0 y_0 \qquad \mu_0 = \|T\|$$

Thus $\lambda_0 = 1/\mu_0$ is a characteristic value for the Sturm-Liouville problem, and y_0 is the associated characteristic function. Let us now find all the characteristic values and all the characteristic functions. Let e_1 be the set of all continuous functions y having the property that $(y, y_0) = 0$. If y_1 and y_2 are in e_1, and if c_1 and c_2 are constants, then $c_1 y_1 + c_2 y_2$ are in e_1; and if y is in e_1, then

$$(Ty, y_0) = (y, Ty_0) = (y, \mu_0 y_0) = \mu_0 (y, y_0) = 0$$

and so Ty is in e_1. And if $\{y_n\}$ is a sequence in e_1, and y is such that

$$\lim_{n \to \infty} \|y - y_n\| = 0$$

then

$$|(y, y_0)| = |(y - y_n, y_0) + (y_n, y_0)| = |(y - y_n, y_0)| \leq \|y - y_n\|$$

and so y is in e_1.

Let $\|T\|_1$ be the least upper bound of the numbers $\|Ty\|$, where $\|y\| = 1$ and y is in e_1. Clearly,

$$\|T\|_1 \leq \|T\|$$

and by an argument similar to the one above, the least upper bound of the numbers $|(Ty, y)|$, where $\|y\| = 1$ and y is in e_1, is precisely $\|T\|_1$. Then, reapplying Theorem 3, we obtain a function y_1 in e_1 for which

$$\|y_1\| = 1 \qquad \text{and} \qquad Ty_1 = \mu_1 y_1 \qquad \mu_1 = \|T\|_1$$

Of course, in the above discussion we have assumed that $\|T\|_1 \neq 0$. Otherwise, as we shall see, the only characteristic function would be y_0. Having obtained y_0 and y_1, we can continue this process step by step. The construction is summarized in the following theorem.

Theorem 4 There exists a sequence $\{\mu_n\}$, finite or infinite, and a sequence of functions $\{y_n\}$ such that

$$\mu_0 \geq \mu_1 \geq \cdots \geq \mu_n \geq \cdots > 0$$
$$(y_n, y_m) = \delta_{nm}$$
$$Ty_n = \mu_n y_n$$

and μ_{n+1} is the least upper bound of the set of numbers $\|Ty\|$, where

$$\|y\| = 1 \qquad \text{and} \qquad (y,y_0) = \cdots = (y,y_n) = 0$$

The sequence is finite in case there is an n_0 such that $\mu_{n_0+1} = 0$.

Theorem 5   The sequence $\{\mu_n\}$ does not terminate, and $\lim \mu_n = 0$.   For any continuous y,

$$\lim_{N \to \infty} \left\| Ty - \sum_{n=0}^{N} \mu_n(y,y_n)y_n \right\| = 0$$

Proof   We first show that if the sequence does not terminate, then $\lim \mu_n = 0$.   Otherwise there would be an $\epsilon > 0$ such that $\mu_n > \epsilon$. Therefore the sequence of functions $\{y_n/\mu_n\}$ has the property that

$$\left\| \frac{y_n}{\mu_n} \right\| < \frac{1}{\epsilon}$$

Therefore $T(y_n/\mu_n) = y_n$ has a uniformly convergent subsequence $\{y_{n_j}\}$.   Since $\{y_{n_j}\}$ is uniformly convergent, $\|y_{n_j} - y_{n_j+1}\| \to 0$. But because $(y_{n_j},y_{n_j+1}) = 0$, $\|y_{n_j+1} - y_{n_j}\| = \sqrt{2}$, which is a contradiction.

If the sequence terminates, for $n = n_0$ let

$$z_{n_0} = y - \sum_{n=0}^{n_0} (y,y_n)y_n$$

$$(z_{n_0},y_j) = 0 \qquad \text{for } j = 0, 1, \ldots, N$$

Therefore $\|Tz_{n_0}\| = 0$, because $\mu_{n_0+1} = 0$.   Then

$$Ty = \sum_{n=0}^{n_0} \mu_n(y,y_0)y_n$$

If the sequence does not terminate, let

$$z_N = y - \sum_{n=0}^{N} (y,y_n)y_n$$

Again,

$$(z_N,y_n) = 0 \qquad \text{for } n = 0, 1, \ldots, N$$

and so

$$\|Tz_N\| \le \mu_{N+1}\|z_N\|$$

However,

$$\|z_N\|^2 = \|y\|^2 - \sum_{n=0}^{N} |(y \ y_n)|^2 \le \|y\|^2$$

Thus

$$\|Tz_N\| \leq \mu_{N+1}\|y\|$$

Because $\lim \mu_n = 0$, this shows that

$$\lim_{N \to \infty} \left\| Ty - \sum_{n=0}^{N} \mu_n(y,y_n)y_n \right\| = \lim_{N \to \infty} \|Tz_N\| = 0$$

It remains to show that the sequence does not terminate. Suppose that g is a function which is continuous, has continuous first and second derivatives, and satisfies the conditions (B) or (P), depending upon which problem we are considering. Then, if f is defined by the equality

$$f = - \frac{d^2g}{dx^2} + q(x)g$$

we have $g = Tf$. If the sequence $\{\mu_n\}$ terminates at $n = n_0$, then

$$g = Tf = \sum_{n=0}^{n_0} \mu_n(f,y_n)y_n = \sum_{n=0}^{n_0} (Tf,y_n)y_n = \sum_{n=0}^{n_0} (g,y_n)y_n$$

Now, if $k \geq 2$ and $r \geq 2$, the functions

$$g(x) = (x - a)^k(x - b)^r$$

have continuous second derivatives and satisfy conditions (B) and (P). If every such function could be represented as a linear combination of $y_0, y_1, \ldots, y_{n_0}$, there would be at most $n_0 + 1$ linearly independent functions of this form. But because a polynomial of degree $d + 1$ is linearly independent of any set of polynomials of degree d or smaller, a contradiction has been reached. Thus the sequence $\{\mu_n\}$ does not terminate.

For each μ_n in the sequence we have obtained, the number $\lambda_n = 1/\mu_n$ is a characteristic value of the Sturm-Liouville problem, and the function y_n is the associated characteristic function. The question can now be asked: Does the sequence of functions $\{y_n\}$ contain every characteristic function? Suppose that there were a characteristic function y, with characteristic value $\lambda = 1/\mu$, which was not among the functions $\{y_n\}$. From the fact that $Ty = \mu y$ and $Ty_n = \mu_n y_n$ we have

$$\mu_n(y_n,y) = (Ty_n,y) = (y_n,Ty) = \mu(y_n,y)$$

Thus, unless $\mu_n = \mu$, $(y_n,y) = 0$. There are only a finite number of such μ_n, because $\lim \mu_n = 0$ [in fact, as we have already seen, there is, at most one such μ_n for (B) and at most two for (P), but the question we are considering can be studied entirely in terms of T, without regard to the Sturm-Liouville problem from which it arose]. Now let

$$\mu_{n_0} = \mu_{n_0+1} = \cdot \cdot \cdot = \mu_{n_0+p} = \mu$$

be a listing of all the elements of the sequence which are equal to μ. Then, from Theorem 5, since $(y_n,y) = 0$ if n does not satisfy $n_0 \leq n \leq n_0 - p$, we have

$$y = \frac{1}{\mu} Ty = \frac{1}{\mu} \sum_{n=n_0}^{n_0+p} \mu(y,y_n)y_n = \sum_{n=n_0}^{n_0+p} (y,y_n)y_n$$

Thus we see that if y is not identically zero, it is a linear combination of the characteristic functions already found. Therefore the only characteristic functions not listed among the $\{y_n\}$ are those characteristic functions associated with a characteristic value for which more than one characteristic function exists, and in this case, the characteristic function is a linear combination of the characteristic functions already listed. We can now state the *fundamental theorem of the Sturm-Liouville problem.*

Theorem 6   Let the Sturm-Liouville problem, with min $q(x) > 0$, and conditions (B) or (P) be given. Then the characteristic values are a sequence of positive numbers

$$\{\lambda_n\}, \lambda_0 \leq \lambda, \leq \cdot \cdot \cdot \leq \lambda_n < \cdot \cdot \cdot$$

and $\lim \lambda_n = \infty$. The set of characteristic functions $\{y_n\}$ is complete in the class of functions which have continuous second derivatives and which satisfy the conditions (B) or (P); that is, for such a function f,

$$\lim_{N \to \infty} \left\| f - \sum_{n=0}^{N} (f,y_n)y_n \right\| = 0$$

or, equivalently,

$$\sum_{n=0}^{\infty} |(f,y_n)|^2 = \|f\|^2$$

If g is continuous on the interval $a \leq x \leq b$, the differential equation

$$\frac{-d^2y}{dx^2} + q(x)y = g$$

has a unique solution y, and

$$\lim_{N \to \infty} \left\| y - \sum_{n=0}^{N} \frac{1}{\lambda_n} (g,y_n)y_n \right\| = 0$$

If the generalized Fourier series

$$\sum_{n=0}^{\infty} \frac{1}{\lambda_n} (g,y_n)y_n$$

converges uniformly, its value is y.

Proof The statement about the characteristic values is a consequence of the fact that $\lambda_n = 1/\mu_n$. If f is as described, then

$$f = Ty$$

where

$$y = -\frac{d^2f}{dx^2} + q(x)f$$

But then

$$\left\| f - \sum_{n=0}^{N} (f,y_n)y_n \right\| = \left\| Ty - \sum_{n=0}^{N} (Ty,y_n)y_n \right\|$$

$$= \left\| Ty - \sum_{n=0}^{N} (y,Ty_n)y_n \right\|$$

$$= \left\| Ty - \sum_{n=0}^{N} \mu_n(y,y_n)y_n \right\|$$

and Theorem 5 shows that this last term has the limit zero as $N \to \infty$.

As for the differential equation, we have already seen that the unique solution is $y = Tg$. But then the statement above is just a restatement of Theorem 5, with $\mu_n = 1/\lambda_n$. Finally, if

$$\sum_{n=0}^{\infty} \frac{1}{\lambda_n} (g,y_n)y_n = y^*$$

the convergence being uniform, then

$$\int_a^b |y - y^*|^2 \, dx = \lim_{N \to \infty} \int_a^b \left| y - \sum_{n=0}^{N} \frac{1}{\lambda_n} (g,y_n)y_n \right|^2 dx = 0$$

and so $|y - y^*|$ must be identically zero.

Problem If $G(x,\xi)$ is the function arising in a Sturm-Liouville problem for which the problem is equivalent to

$$y = \int_a^b G(x,\xi)q(\xi) \, d\xi$$

and if $\{y_n\}$ and $\{\lambda_n\}$ are the characteristic functions and characteristic values, show that

$$\lim_{N \to \infty} \int_a^b \left| G(x,\xi) - \sum_{n=1}^{N} \frac{y_n(x)y_n(\xi)}{\lambda_n} \right|^2 dx = 0$$

APPROXIMATE VALUES OF λ_n AND y_n FOR LARGE VALUES OF n

We shall now show that as n becomes large, the values of λ_n can be given, approximately, as n^2, and that the characteristic functions y_n are, approximately, trigonometric functions of a particularly simple form. The precise meaning of this statement will become clear as we proceed. We shall deal only with conditions (B); the argument for conditions (P) is similar.

Let $y(x,p)$ be the solution of the equation

$$\frac{-d^2y}{dx^2} + q(x)y = p^2 y \qquad \text{for } a < x < b$$

where

$$y(a,p) = \cos \alpha \qquad y'(a,p) = \sin \alpha$$

Then $y(x,p)$ satisfies the first of the two conditions in (B). In order that $y(x,p)$ be a characteristic function, it is necessary and sufficient that

$$\sin \beta \, y(b,p) - \cos \beta \, y'(b,p) = 0$$

In this case, p^2 will be a characteristic value. Conversely, if p^2 is a characteristic value, the associated characteristic function must be a constant multiple of $y(x,p)$; thus the above equality can be regarded as the equation which determines the characteristic values.

Now, because $y(x,p)$ is uniquely determined by the differential equation and the initial values,

$$y(x,p) = \cos \alpha \cos p(x - a) + \sin \alpha \, \frac{\sin p(x - a)}{p}$$
$$+ \int_a^x \frac{\sin p(x - \zeta)}{p} q(\zeta)y(\zeta,p) \, d\zeta$$

For $p > 1$,

$$|y(x,p)| \le \sqrt{\cos^2 \alpha + \frac{1}{p^2} \sin^2 \alpha} + \int_a^x \frac{1}{p} q(\zeta)|y(\zeta,p)| \, d\zeta$$
$$\le 1 + \int_a^x \frac{1}{p} q(\zeta)|y(\zeta,p)| \, d\zeta$$

Let $H(x)$ be the right-hand member of the inequality. Then $H(a) = 1$, and

$$\frac{dH}{dx} = \frac{1}{p} q(x)|y(x,p)| < \frac{1}{p} q(x) H(x)$$

If we multiply this last inequality by the positive function

$$\exp\left[-\frac{1}{p} \int_a^x q(\zeta)\, d\zeta \right]$$

we obtain

$$\frac{d}{dx}\left\{ H(x) \exp\left[-\frac{1}{p} \int_a^x a(\zeta)\, d\zeta \right] \right\} \leq 0$$

and integration from a to x yields

$$H(x) \leq \exp\left[\frac{1}{p} \int_a^x q(\zeta)\, d\zeta \right]$$

Therefore, for $a < x < b$,

$$|y(x,p)| \leq \exp\left[\frac{1}{p} \int_a^b q(b)\, d\zeta \right]$$

Thus if $A > 1$, there exists a $p_0 > 0$ such that if $p > p_0$, then $y(x,p) \leq A$ for all x in $a < x < b$.

Let us designate $\partial/\partial p\, y(x,p)$ by the symbol $y_p(x,p)$. Because $y(a,p) = \cos \alpha$,

$$\frac{dy}{dx}(a,p) = \sin \alpha \qquad \text{for all } p$$

$$y_p(a,p) = \frac{dy_p}{dx}(a,p) = 0$$

Furthermore, $y_p(x,p)$ satisfies the differential equation

$$\frac{d^2 y_p}{dx^2} + p^2 y_p = q(x)y_p - 2py$$

Therefore

$$y_p(x,p) = \int_a^x \frac{\sin p(x - \zeta)}{p} \left[q(\zeta)yp(\zeta,p) - 2py(\zeta,p) \right]$$

Thus

$$|y_p(x,p)| \leq 2 \int_a^x |y(\zeta,p)|\, d\zeta + \int_a^x \frac{1}{p} q(\zeta)|y_p(\zeta,p)|\, d\zeta$$

If $p > p_0$,

$$|y(\zeta,p)| < A \qquad \text{and} \qquad |y_p(x,p)| \le 2(b - a)A$$
$$+ \int_a^x \frac{1}{p} q(\zeta)|y_p(\zeta,p)| \, d\zeta$$

By an argument similar to that just given, if $a < x < b$,

$$|y_p(x,p)| < 2(b - a)A \exp\left[\frac{1}{p} \int_a^b q(\zeta) \, d\zeta\right]$$

Thus if $B > 2(b - a)A$, there is a $p_1 > p_0$ such that if $p > p_1$, then $y_p(x,p) < B$ for all x in $a < x < b$.

We shall now deal with several different cases.

Case I: cos α ≠ 0   In this case

$$y(b,p) = \cos \alpha \cos p(b - a) + \frac{C_1(p)}{p}$$

where

$$C_1(p) = \sin \alpha \sin p(b - a) + \int_a^b \sin p(b - \zeta) \, q(\zeta)y(\zeta,p) \, d\zeta$$

If $p > p_1$, then

$$C_1(p) \le \sin \alpha + A \int_a^b |q(\zeta)| \, d\zeta = K$$

and

$$\left|\frac{dC_1}{dp}(p)\right| < (b - a) \sin \alpha + A \int_a^b (b - \zeta)|q(\zeta)| \, d\zeta$$
$$+ B \int_a^b q(\zeta) \, d\zeta = L$$

Now,

$$y'(b,p) = -p \cos \alpha \sin p(b - a) + C_2(p)$$

where

$$C_2(p) = \sin \alpha \cos p(b - a) + \int_a^b \cos p(b - \zeta) \, q(\zeta)y(\zeta,p) \, d\zeta$$

If $p > p_1$,

$$|C_2(p)| \le K \qquad \text{and} \qquad |dC_2/dp| \le L$$

Then

$$\sin \beta \, y(b,p) - \cos \beta \, y'(b,p) = \sin \beta \left[\cos \alpha \cos p(b - a) + \frac{C_1(p)}{p}\right]$$
$$= \cos \beta \left[-p \cos \alpha \sin p(b - a) + C_2(p)\right]$$

If $\cos \beta \neq 0$, the values of p for which p^2 is a characteristic value are the solutions of the equation

$$\sin p(b - a) = \frac{1}{p} \left[\frac{C_2(p)}{\cos \beta} - \frac{\tan \beta}{p \cos \alpha} C_1(p) \right.$$

$$\left. - \tan \beta \cos p(b - a) \right] = F(p)$$

If $p > p_1$, the right-hand member of this equation is bounded by the quantity

$$\frac{1}{p} \left[K \left(\frac{1}{\cos \beta} + \frac{1}{p} \left| \frac{\tan \beta}{\cos \alpha} \right| \right) + |\tan \beta| \right]$$

and its derivative is bounded by the quantity

$$\frac{1}{p} \left[K \left(\frac{1}{\cos \beta} + \frac{1}{p} \left| \frac{\tan \beta}{\cos \alpha} \right| \right) + |\tan \beta| \right]$$

$$+ \frac{1}{p} L \left(\frac{1}{\cos \beta} + \frac{\tan \beta}{\cos \alpha} \frac{1}{p} \right) + \left| \frac{\tan \beta}{\cos \alpha} \right| \frac{K}{p}$$

Therefore both $F(p)$ and dF/dp approach zero as $p \to \infty$.

Assertion   Given ϵ, $0 < \epsilon < \frac{1}{2}$, there exists a $\bar{p} > 0$ such that for $n\pi/(b - a) > \bar{p}$ the interval

$$\frac{(n - \epsilon)\pi}{b - a} < p < \frac{(n + \epsilon)\pi}{b - a}$$

contains exactly one root of the equation

$$\sin p(b - a) = F(p)$$

and no other values of $p > \bar{p}$ are roots of this equation.

Proof   Let p_2 be such that if $p > p_2$, $F(p) < \sin \epsilon\pi$.   Let

$$p_n^+ = \frac{(n + \epsilon)\pi}{b - a} \qquad p_n^- = \frac{(n - \epsilon)\pi}{b - a}$$

In the interval $p_n^+ < p < p_{n+1}^-$, the function $\sin p(b - a)$ has its minimum value at the end points, and at the end points its value is $\sin \epsilon\pi$.   Therefore if p lies in this interval, and $n\pi/(b - a) > p_2$, then $\sin p(b - a) - F(p) = 0$ has no roots.   The two numbers $\sin p_n^+(b - a)$ and $\sin p_n^-(b - a)$ have the same absolute value, $\sin \epsilon\pi$, but are opposite in sign.   Therefore, if $n\pi/(b - a) > p_2 + \pi/(b - a)$, the function $\sin p(b - a) - F(p)$ has different signs at the two points p_n^- and p_n^+.   Hence there is a $\tilde{p}$, $p_n^- < \tilde{p} < p_n^+$, for which this function is zero.   If there were two such roots, by

Rolle's theorem, there would be a p^*, $p_n^- < p^* < p_n^+$, for which the derivative vanished; that is,

$$(b - a) \cos p^*(b - a) - F'(p^*) = 0$$

But on this interval the minimum value of $\cos p(b - a)$ is assumed at the end points, where its value is $\cos \epsilon\pi$. There exists a $p_3 > 0$ such that if $p > p_3$, then

$$F'(p) < (b - a) \cos \epsilon\pi$$

If $n\pi/(b - a) > p_3 + \pi/(b - a)$, then $(b - a) \cos p^*(b - a) - F'(p^*)$ cannot be zero for $p_n^- < p^* < p_n^+$. Therefore, if $\bar{p}$ is the larger of the numbers $p_2 + \pi/(b - a)$ and $p_3 + \pi/(b - a)$, the assertion is true.

If $\cos \beta = 0$, instead of the equation

$$\sin p(b - a) = F(p)$$

we must deal with the equation

$$\cos p(b - a) = -\frac{C_1(p)}{\cos d\, p}$$

which is treated in a similar fashion. Here, however, the appropriate interval is

$$\frac{(n + \frac{1}{2} - \epsilon)\pi}{b - a} < p < \frac{(n + \frac{1}{2} + \epsilon)\pi}{b - a}$$

Case II: cos α = 0   In this case

$$y(x,p) = \sin \alpha\, \frac{\sin p(x - a)}{p} + \int_a^x \frac{\sin p(x - \zeta)}{p}\, q(\zeta)y(\zeta,p)\, d\zeta$$

and

$$y'(x,p) = \sin \alpha \cos p(x - a) + \int_a^x \cos p(x - \zeta)\, q(\zeta)y(\zeta,p)\, d\zeta$$

Then

$$y(b,p) = \sin \alpha\, \frac{\sin p(b - a)}{p} + \frac{D_1(p)}{p^2}$$

$$y'(b,p) = \sin \alpha \cos p(b - a) + \frac{D_2(p)}{p}$$

where $D_1(p)$ and $D_2(p)$ are, for sufficiently large p, bounded functions with bounded derivatives. The equation which determines the values of

p for which p^2 is a characteristic value is then

$$\sin \beta \left[\sin \alpha \, \frac{\sin p(b - a)}{p} + \frac{D_1(p)}{p^2} \right]$$

$$= \cos \beta \left[\sin \alpha \cos p(b - a) + \frac{D_2(p)}{p} \right]$$

For $\cos \beta \neq 0$, the equation to be dealt with is

$$\cos p(b - a) = F_1(p)$$

where $F_1(p)$ and $F_1'(p)$ both approach zero as $p \to \infty$, and for $\cos \beta = 0$, the appropriate equation is

$$\sin p(b - a) = F_1(p)$$

The results are similar to those in case I.

Theorem 7 For the Sturm-Liouville problem with conditions (B), the characteristic values $\lambda_n = p_n{}^2$ satisfy the condition

$$\lim_{n \to \infty} \frac{\lambda_n (b - a)^2}{\pi^2 n^2} = 1$$

The characteristic functions y_n, except for a constant factor, are of the form

$$\cos p_n(x - a) + \frac{B}{n} \qquad \text{for } \cos \alpha \neq 0$$

$$\sin p_n(x - a) + \frac{C}{n} \qquad \text{for } \cos \alpha = 0$$

where B and C are bounded functions. For arbitrary continuous g the generalized Fourier series

$$\sum_{n=0}^{\infty} \frac{1}{\lambda_n} (g, y_n) y_n$$

converges uniformly and absolutely.

Proof All the statements of the theorem except the last have already been proved. Now, if

$$y_n = \alpha_n \left[\cos p_n(x - a) + \frac{B}{n} \right]$$

then

$$\int_a^b |(y_n)|^2 \, dx = \alpha_n{}^2 \left[\frac{b - a}{2} + u(n) \right]$$

where $u(n)$ is bounded, and

$$\alpha_n{}^2 = \frac{1}{[(b-a)/2] + u(n)}$$

Hence α_n is bounded as $n \to \infty$. Therefore there exists a constant $\Lambda > 0$ such that $|y_n(x)| < \Lambda$ for all n and for all x, $a < x < b$. Then, by the Cauchy-Schwartz inequality,

$$\sum_{n=0}^{N} \frac{1}{\lambda_n} |(g,y_n)| \, |y_n| \le \Lambda \left(\sum_{n=0}^{N} \frac{1}{\lambda_n{}^2} \right)^{\!1/2} \left[\sum_{n=0}^{N} |(g,y_n)|^2 \right]^{1/2}$$

$$\le \Lambda \, \|g\| \left(\sum_{n=0}^{N} \frac{1}{\lambda^2} \right)^{\!1/2}$$

Because

$$\lim_{n \to \infty} \frac{\lambda_n (b-a)^2}{\pi^2 n^2} = 1$$

$\displaystyle\sum_{n=0}^{\infty} 1/\lambda_n{}^2$ is a convergent series, and therefore $\displaystyle\sum_{n=0}^{\infty} (1/\lambda_n)|(g,y_n)| \, |y_n|$ converges uniformly.

9
The Bessel Functions

THE BESSEL EQUATION

The differential equation

$$r^2 \frac{d^2y}{dr^2} + r \frac{dy}{dr} + (r^2 - \nu^2)y = 0$$

is called *the Bessel equation* of order ν. We shall assume that $\nu \geq 0$. As a differential equation, it has a regular singular point at $r = 0$. If we look for a solution of the form

$$y = x^\alpha \sum_{n=0}^{\infty} A_n x^n \qquad a_0 = 1$$

the indicial equation is

$$\alpha^2 - \nu^2 = 0$$

and the recursion relations are

$$
a_n = \begin{cases}
- \dfrac{a_{n-2}}{n(n + 2\nu)} & \text{for } \alpha = \nu \\[3mm]
\dfrac{a_{n-2}}{n(n - 2\nu)} & \text{for } \alpha = -\nu
\end{cases}
$$

If 2ν is not an even integer, we obtain two solutions of the form

$$
r^\nu(1 + E) \qquad r^{-\nu}(1 + E)
$$

where E denotes a power series in powers of r^2. If 2ν is an even integer, or, equivalently, if ν is an integer, the second recursion relation does not yield a solution; in this case we let

$$
y_1 = r^\nu(1 + E)
$$

be the solution arising from the first recursion relation, and we seek a solution of the form

$$
y_2 = uy
$$

Substituting y_2 into the differential equation, and remembering that y_1 is a solution of the equation, we find that y_2 will be a solution if u satisfies

$$
\frac{d^2u}{dr^2} + \left(\frac{2\, dy_1/dr}{y_1} + \frac{1}{r} \right) \frac{du}{dr} = 0
$$

or, equivalently,

$$
\frac{d}{dr} \log \left(y_1{}^2 \frac{du}{dr} \right) = 0
$$

But then

$$
\frac{du}{dr} = \frac{C}{ry_1{}^2}
$$

where C is a constant. The right-hand member of this equation is of the form

$$
\frac{C}{r^{2\nu+1}} [1 + E]
$$

Therefore u is of the form

$$
u = C_1 \log r + \frac{C_2}{r^{2\nu}} [1 + E]
$$

where C_1 and C_2 constant. Then y_2 is of the form

$$
y_2 = C_1(\log r)y_1 + C_2 r^{-\nu}[1 + E]
$$

The solution $y_1 = r^\nu(1 + E)$ is called the *Bessel function* of order ν, and the other solution, y_2, is called the *Neumann function* of order ν. These functions have been studied extensively, even for complex ν and complex r. The Bessel functions are usually designated by the symbol $J_\nu(r)$ and the Neumann functions by either $N_\nu(r)$ or $I_\nu(r)$. It is conventional when using these symbols to multiply the functions y_1 and y_2 by certain constants, so that their leading coefficients have certain conventional values; these are of no significance to us here, except for the observation that $J_0(r)$ is, in fact, of the form $1 + E$.

THE ZEROS OF $J_0(r)$ AND THE RELATED STURM-LIOUVILLE PROBLEM

The problem we set for ourselves is to find those values of r for which $J_0(r) = 0$. Let k be such a value of r, and let

$$y(x) = J_0(kx)$$

Then $y(x)$ satisfies the differential equation

$$x \frac{d^2y}{dx^2} + \frac{dy}{dx} + k^2xy - 0$$

In addition, $y(0) = 1$ and $y(1) = 0$.

Theorem 1 If $J_0(k) = 0$, then k is real.

Proof From the preceding remarks, we have

$$\frac{d}{dx}\left(x \frac{dy}{dx} \right) + k^2xy = 0$$

and

$$\frac{d}{dx}\left(x \frac{d\bar{y}}{dx} \right) + \bar{k}^2x\bar{y} = 0$$

If we multiply the first of these equations by $\bar{y}$ and the second by y, and both add and subtract the two resulting equations, we obtain

$$\bar{y} \frac{d}{dx}\left(x \frac{dy}{dx} \right) - y \frac{d}{dx}\left(x \frac{d\bar{y}}{dx} \right) + (k^2 - \bar{k}^2)x|y|^2 = 0$$

and

$$\bar{y} \frac{d}{dx}\left(x \frac{dy}{dx} \right) + y \frac{d}{dx}\left(x \frac{d\bar{y}}{dx} \right) + (k^2 + \bar{k}^2)x|y|^2 = 0$$

If we integrate both sides of both of these equations from 0 to 1, making use of the fact that $y(1) = \bar{y}(1) = 0$, we obtain

$$(k^2 - \bar{k}^2) \int_0^1 x|y|^2 \, dx = 0$$

and

$$(k^2 + \bar{k}^2) \int_0^1 x|y|^2 \, dx = 2 \int_0^1 x \left| \frac{dy}{dx} \right|^2 d$$

Therefore $k^2 - \bar{k}^2 = 0$, and so k^2 is real, and $k^2 + \bar{k}^2 > 0$, and so k^2 is nonnegative. Therefore k is real.

Now, suppose that $y(x)$ satisfies the above equation. Then if $v(r) = y(r/k)$, we have

$$r\frac{d^2v}{dr^2} + \frac{dv}{dr} + rv = 0$$
$$v(0) = 1 \qquad v(k) = 0$$

Because v is a solution of the Bessel equation, v is a linear combination of $J_0(r)$ and $N_0(r)$. But $N_0(r)$ has a logarithmic singularity at $r = 0$, and so v must be a multiple of $J_0(r)$. Because $v(0) = 1$, $v = J_0(r)$, and so $J_0(k) = 0$. Therefore the problem of finding those values of k for which $J_0(k) = 0$ is equivalent to finding values of k for which

$$\frac{d}{dx}\left(x\frac{dy}{dx} \right) + k^2 xy = 0 \qquad \text{for } 0 < x < 1$$

has a solution $y(x)$ for which $y(0) = 1$ and $y(1) = 0$. Let $y(x) = x^{-\frac{1}{2}}f(x)$. Then

$$\frac{dy}{dx} = x^{-\frac{1}{2}} \frac{df}{dx} - \frac{1}{2} x^{-\frac{1}{2}} f$$
$$x^{\frac{1}{2}} \frac{d^2f}{dx^2} + \frac{1}{4} x^{-\frac{3}{2}} f + k^2 x^{\frac{1}{2}} f = 0$$

or

$$-\frac{d^2f}{dx^2} - \frac{1}{4x^2} f = k^2 f$$

The problem can now be stated in the form of a theorem.

Theorem 2 $J_0(k) = 0$ if and only if

$$J_0(kx) = Cx^{-\frac{1}{2}}f(x)$$

where C is a constant, f satisfies the equation

$$-\frac{d^2f}{dx^2} - \frac{1}{4x^2} f = \lambda f \qquad \lambda = k^2$$

$f(1) = 0$, and $x^{-\frac{1}{2}}f(x)$ is bounded.

Proof   We have just seen that if, instead of saying that $x^{-\frac{1}{2}}f(x)$ is bounded, we had said

$$\lim_{x \to 0} x^{-\frac{1}{2}}f(x) = 1$$

the theorem would be true for $C = 1$.   Any f which satisfies the conditions of the theorem can be converted to an f for which

$$\lim_{x \to 0} x^{-\frac{1}{2}}f(x) = 1$$

by multiplying by a constant, unless

$$\lim_{x \to 0} x^{-\frac{1}{2}}f(x) = 0$$

But the indicial equation for the differential equation of the theorem has roots $\pm\frac{1}{2}$.   Therefore any solution of this equation is of the form

$$f = [C_1 x^{\frac{1}{2}} + C_2 x^{-\frac{1}{2}}][1 + E]$$

If

$$\lim_{x \to 0} x^{-\frac{1}{2}}f(x) = 0$$

both C_1 and C_2 are zero.

The problem of finding a value of λ and a function f such that

$$-\frac{d^2 f}{dx^2} - \frac{1}{4x^2}f = \lambda f$$

$f(1) = 0$, and $x^{-\frac{1}{2}}f(x)$ is bounded resembles very closely the problem of finding the characteristic values and characteristic functions for the Sturm-Liouville problem.   Truly, $q(x) = -1/4x^2$ is neither continuous on $0 < x < 1$, nor is it positive.   The boundary condition $f(1) = 0$ is of the type of the second condition of (B), but the other condition of f is quite different.   This problem is called a *singular Sturm-Liouville problem*.   The numbers λ continue to be called characteristic values, and the associated functions f continue to be called characteristic functions. The general theory of singular Sturm-Liouville problems shares some, but not all, of the features of the problem we studied in the preceding chapter.   However, in the case at hand the main facts are identical.

Theorem 3   The characteristic values λ and the characteristic functions f of the singular Sturm-Liouville problem,

$$-\frac{d^2 f}{dx^2} - \frac{1}{4x^2}f = \lambda f \qquad \text{for } 0 < x \le 1$$

$f(1) = 0$, and $x^{-\frac{1}{2}}f(x)$ bounded, form, respectively, an increasing sequence $\{\lambda_n\}$ of positive numbers, where $\lim_{n \to \infty} \lambda_n = \infty$, and an orthonormal sequence of functions $\{f_n\}$ which is complete in the class of functions g which have continuous second derivatives and satisfy the conditions $g(1) = 0$ and $x^{-\frac{1}{2}}g$ bounded. Furthermore, if g is continuous on $0 \leq x \leq 1$ and

$$-\frac{d^2f}{dx^2} - \frac{1}{4x^2} f = g$$

then

$$\lim_{N \to \infty} \left\| \sum_{n=0}^{N} \frac{1}{\lambda_n} (g,f_n)f_n - f \right\| = 0$$

Proof We have already seen that if we write $\lambda = k^2$, finding the characteristic values is equivalent to finding the roots k of the equation $J_0(k) = 0$. Therefore the characteristic values are nonnegative, and in fact, because $J_0(0) = 1$, they are positive. All that remains is to find the function $G(x,\zeta)$ such that

$$f = Tg = \int_0^1 G(x,\zeta)g(\zeta) \, d\zeta$$

is the solution of the above equation. The Sturm-Liouville theorem of Chap. 8 followed from the fact that $G(x,\zeta) = \overline{G(\zeta,x)}$ and $G(x,\zeta)$ is bounded on the rectangle $0 < x < 1$ and $0 < \zeta < 1$. But it is easy to verify that for this problem

$$G(x,\zeta) = \begin{cases} -x^{\frac{1}{2}}\zeta^{\frac{1}{2}} \log \zeta & \text{for } x < \zeta \\ -x^{\frac{1}{2}}\zeta^{\frac{1}{2}} \log x & \text{for } \zeta < x \end{cases}$$

and that this function has the desired properties.

Problem

(a) Show that if $u(r)$ is a function defined on $0 \leq r \leq a$, $u(r)$ is bounded as $r \to a$, and $u(a) = 1$, then in order that k^2 be a number such that

$$r^2 \frac{d^2u}{dr^2} + r \frac{du}{dr} + (k^2r^2 - n^2)u = 0 \qquad \text{for } n > 0$$

$u(r)$ must satisfy the equation

$$u(r) = -k^2 \int_0^a G(r,r_1)u(r_1)r_1 \, dr_1$$

where

$$G(r,r_1) = \begin{cases} \dfrac{1}{2n}\dfrac{(r_1 r)^n}{a^{2n}} - \dfrac{1}{2n}\left(\dfrac{r_1}{r}\right)^n & \text{for } r_1 < r \\[2ex] \dfrac{1}{2n}\dfrac{(r_1 r)^n}{a^{2n}} - \dfrac{1}{2n}\left(\dfrac{r_1}{r}\right)^n & \text{for } r_1 > r \end{cases}$$

(b) By making the change of variables $r^2 = x$ and $r_1{}^2 = \zeta$, show that this integral equation is one to which the Sturm-Liouville theory is applicable.

(c) Find all values of k for which there exists a function $\omega(r,\theta)$ of the form

$$\omega(r,\theta) = u(r)v(\theta)$$

which is continuous and bounded in the circle $r \le a$, is zero at $r = a$, and satisfies the equation

$$\frac{\partial^2 \omega}{\partial r^2} + \frac{1}{r}\frac{\partial \omega}{\partial r} + \frac{1}{r^2}\frac{\partial^2 \omega}{\partial \theta^2} + k^2\omega = 0$$

Theorem 3 can be rewritten in terms of the Bessel functions in the following fashion.

Theorem 4   The positive zeros of the Bessel function form an increasing sequence $\{k_n\}$, where $\lim\limits_{n\to\infty} k_n = \infty$, and the functions

$$\phi_n(r) = \frac{r^{1/2}J_0(k_n r)}{\left\{\displaystyle\int_0^1 r[J_0(k_n r)]^2\, dr\right\}^{1/2}}$$

form an orthonormal sequence $\{\phi_n\}$ which is complete in the class of functions g which have continuous second derivatives and for which $g(1) = 0$ and $r^{-1/2}g(r)$ is bounded.   That is, the Fourier-Bessel expansion

$$g \sim \sum_{n=0}^{\infty} (g,\phi_n)\phi_n$$

or, equivalently,

$$g \sim \sum_{n=0}^{\infty} \frac{\displaystyle\int_0^1 g(r)J_0(k_n r)\sqrt{r}\, dr}{\displaystyle\int_0^1 r[J_0(k_n r)]^2\, dr} \sqrt{r}\, J_0(k_n r)$$

is such that

$$\lim_{N\to\infty}\left\| g - \sum_{n=0}^{N} \frac{1}{\lambda_n}(g,f_n)f_n \right\| = 0$$

If h is a continuous function, the differential equation

$$\frac{d^2y}{dr^2} + \frac{1}{r}\frac{dy}{dr} = -h$$

has a unique solution for which $y(1) = 0$ and $y(r)$ is bounded, and

$$\lim_{N \to \infty} \left\| g - \sum_{n=0}^{N} \frac{1}{k_n{}^2} \frac{\int_0^1 h(r)J_0(k_nr)r\,dr}{\int_0^1 rJ_0(k_nr)^2\,dr} J_0(k_nr) \right\| = 0$$

ESTIMATION OF k_n FOR LARGE n

We now repeat, in spirit, the material of the last chapter. We shall show that for large n, k_n is approximately $n + 3\pi/4$ and $J_0(k_nr)$ is approximately $\sqrt{2/\pi k_n r} \cos(k_n r - \pi/4)$. Once this has been shown, the convergence of the series

$$\sum_{n=0}^{N} \frac{1}{k_n{}^2} \frac{\int_0^1 h(r)J_0(k_nr)r\,dr}{\int_0^1 J_0(kr)^2r\,dr} J_0(k_nr)$$

absolutely and uniformly, is assured. However, because we are dealing with the singular problem, the methods we use are quite different.

Theorem 5

$$J_0(r) = \frac{1}{2\pi} \int_{-\pi}^{\pi} \cos(r\cos\theta)\,d\theta$$

Proof   Let

$$F(r) = \frac{1}{2\pi} \int_{-\pi}^{\pi} \cos(r\cos\theta)\,d\theta$$

Then

$$\frac{dF}{dr} = -\frac{1}{2\pi} \int_{-\pi}^{\pi} \sin(r\cos\theta)\cos\theta\,d\theta$$

and

$$\frac{d^2F}{dr^2} = -\frac{1}{2\pi} \int_{-\pi}^{\pi} \cos(r\cos\theta)\cos^2\theta\,d\theta$$

Therefore

$$\frac{d^2F}{dr^2} + \frac{1}{r}\frac{dF}{dr} + F = \frac{1}{2\pi} \int_{-\pi}^{\pi} \sin^2\theta\cos(r\cos\theta)\,d\theta$$

$$- \frac{1}{2\pi r} \int_{-\pi}^{\pi} \sin(r\cos\theta)\cos\theta\,d\theta$$

But

$$\frac{1}{2\pi r} \int_{-\pi}^{\pi} \sin(r \cos \theta) \cos \theta \, d\theta = \frac{1}{2\pi r} \int_{-\pi}^{\pi} \sin(r \cos \theta) \, d(\sin \theta)$$

$$= \frac{1}{2\pi} \int_{-\pi}^{\pi} \sin^2 \theta \cos(r \cos \theta) \, d\theta$$

Therefore $F(r)$ satisfies the Bessel equation of order zero, and since $F(0) = 1$, $F(r) = J_0(r)$.

Theorem 6

$$J_0(r) = \sqrt{\frac{2}{\pi r}} \left[\cos\left(\frac{r - \pi}{4}\right) + \frac{B(r)}{r} \right]$$

where $B(r)$ is a function of r which is bounded as $n \to \infty$, as is $B'(r)$.

Proof   From Theorem 5,

$$J_0(r) = \frac{1}{2\pi} \int_{-\pi}^{\pi} \cos(r \cos \theta) \, d\theta = \frac{1}{\pi} \int_{0}^{\pi} \cos(r \cos \theta) \, d\theta$$

$$= \frac{2}{\pi} \int_{0}^{\pi/2} \cos(r \cos \theta) \, d\theta$$

Now,

$$\int_{0}^{\pi/2} \cos(r \cos \theta) \, d\theta = \cos r \int_{0}^{\pi/2} \cos r(1 - \cos \theta) \, d\theta$$

$$+ \sin r \int_{0}^{\pi/2} \sin r(1 - \cos \theta) \, d\theta$$

We shall now concentrate on the integral

$$\int_{0}^{\pi/2} \cos r(1 - \cos \theta) \, d\theta$$

If we let $\varsigma^2 = r(1 - \cos \theta)$, this integral is transformed into

$$\sqrt{\frac{2}{r}} \int_{0}^{\sqrt{r}} \cos \varsigma^2 \left(1 - \frac{\varsigma^2}{2r}\right)^{-\frac{1}{2}} d\varsigma$$

$$= \sqrt{\frac{2}{r}} \left\{ \int_{0}^{\sqrt{r}} \cos \varsigma^2 \, d\varsigma + \int_{0}^{\sqrt{r}} \cos \varsigma^2 \left[\left(1 - \frac{\varsigma^2}{2r}\right)^{-\frac{1}{2}} - 1\right] d\varsigma \right\}$$

We now have two integrals with which to deal.   As to the first, we observe that

$$\frac{d}{dA} \int_{0}^{A} e^{-2Ax} \sin(A^2 - x^2) \, dx$$

$$= \int_{0}^{A} [-2x \sin(A^2 - x^2) + 2A \cos(A^2 - x^2)] e^{-2Ax} \, dx$$

$$= \int_{0}^{A} \frac{d}{dx} \{ - [\cos(A^2 - x^2)] e^{-2Ax} \} \, dx$$

$$= \cos A^2 - e^{-2A^2}$$

Therefore

$$\int_0^A \cos \zeta^2 \, d\zeta = \int_0^A e^{-2\zeta^2} \, d\zeta + \int_0^A e^{-2A\zeta} \sin (A^2 - \zeta^2) \, d\zeta$$

$$= \int_0^\infty e^{-2\zeta^2} \, d\zeta + \int_0^A e^{-2A\zeta} \sin (A^2 - \zeta^2) \, d\zeta - \int_A^\infty e^{-2\zeta^2} \, d\zeta$$

The first integral on the right is precisely equal to $\frac{1}{2} \sqrt{\pi/2}$. The second is no greater in absolute value than

$$\int_0^A e^{-2A\zeta} \, d\zeta = \frac{1 - e^{-2A\zeta}}{2A} \le \frac{1}{2A}$$

The third is equal in absolute value to

$$\frac{1}{2} \int_{A^2}^\infty \frac{e^{-2\zeta}}{\zeta} \, d\zeta \le \frac{1}{2A} \int_{A^2}^\infty e^{-2\zeta} \, d\zeta \le \frac{1}{4A}$$

Therefore

$$\int_0^{\sqrt{r}} \cos \zeta^2 \, d\zeta = \frac{1}{2} \sqrt{\frac{\pi}{2}} + \frac{B_1(r)}{\sqrt{r}}$$

where $B_1(r) \le \frac{3}{4}$. It is elementary to verify that $dB_1(r)/dr$ is bounded.

Now,

$$\int_0^{\sqrt{r}} \cos \zeta^2 \left[\left(1 - \frac{\zeta^2}{2r} \right)^{-\frac{1}{2}} - 1 \right] d\zeta = \int_0^{\sqrt{r}} \left(\sum_{n-1}^\infty a_n \frac{\zeta^{2n}}{2^n r^n} \right) \cos \zeta^2 \, d\zeta$$

where a_n are the binomial coefficients. The series converges uniformly on the interval of integration, and so the integral can be written as

$$\sum_{n=1}^\infty \frac{a_n}{2^n r^n} \int_0^{\sqrt{r}} \zeta^{2n-1} \, d \frac{\sin \zeta^2}{2}$$

$$= \sum_{n=1}^\infty \frac{a_n}{2^n r^n} \left[r^{n-\frac{1}{2}} \frac{\sin r}{2} - \int_0^{\sqrt{r}} (2n - 1)\zeta^{2n-2} \frac{\sin \zeta^2}{2} \, d\zeta \right]$$

The integral here is bounded in absolute value by $\frac{1}{2} r^{n-\frac{1}{2}}$. Therefore

$$\int_0^{\sqrt{r}} \cos \zeta^2 \left[\left(1 - \frac{\zeta^2}{2r} \right)^{-\frac{1}{2}} - 1 \right] d\zeta = \frac{B_2(r)}{\sqrt{r}}$$

where

$$|B_2(r)| \le \sum_{n=1}^\infty \frac{a_n}{2^n} = \sqrt{2} - 1$$

Again, it can be verified that dB_2/dr is bounded.

Therefore

$$\int_0^{\pi/2} \cos r(1 - \cos \theta)\, d\theta = \sqrt{\frac{2}{r}} \left(\frac{1}{2}\sqrt{\frac{\pi}{2}} + \frac{B_1 + B_2}{\sqrt{r}} \right)$$

A similar argument gives

$$\int_0^{\pi/2} \sin r(1 - \cos \theta)\, d\theta = \sqrt{\frac{2}{r}} \left(\frac{1}{2}\sqrt{\frac{\pi}{2}} + \frac{B_3 + B_4}{\sqrt{r}} \right)$$

$$\frac{2}{\pi} \int_0^{\pi/2} \cos (r \cos \theta)\, d\theta$$

$$= \sqrt{\frac{1}{r\pi}} \left[\cos r + \sin r + \frac{\cos r\,(B_1 + B_2)}{\sqrt{r}} \right] + \frac{\sin r\,(B_3 + B_4)}{\sqrt{r}}$$

$$= \sqrt{\frac{2}{\pi r}} \cos \left(r - \frac{\pi}{4} \right) + \frac{B}{\sqrt{r}}$$

Theorem 7 If $0 < \epsilon < \frac{1}{2}$, there exists an $N > 0$ such that for $n > N$ the interval

$$(n + \tfrac{3}{4} - \epsilon)\pi < r < (n + \tfrac{3}{4} + \epsilon)\pi$$

contains exactly one zero of $J_0(r)$, and for $r > N\pi$, $J_0(r)$ has no other zeros.

Proof The proof is so similar to the corresponding theorem in Chap. 8 that we shall not bother to repeat it.

appendix **A**

Continuum Dynamics

KINEMATICS

Points in the three-dimensional euclidean space $E^{(3)}$ will be designated by vectors $\mathbf{s}$, $\mathbf{x}$, . . . , and the components of these vectors with respect to a fixed set of orthogonal coordinate axes in $E^{(3)}$ will be designated as s_j, x_j, . . . , where $j = 1, 2, 3$. We will be dealing with vector-valued or real-valued functions of points in $E^{(3)}$ and of the real variable t, which will be identified with time. For simplicity, these functions will be assumed to have continuous second partial derivatives with respect to the independent variables.

A *continuum* is an open connected set in $E^{(3)}$. If C_0 is a continuum having points labeled by $\mathbf{s}$, let

$$\mathbf{x} = \mathbf{A}(\mathbf{s}, t) \tag{1}$$

be a family of one-to-one mappings of C_0 into $E^{(3)}$. For each t, the image of C_0 is a continuum C_t; we shall assume that $\mathbf{A}(\mathbf{s}, 0) = \mathbf{s}$. For each $\mathbf{s}$ in C_0 the mapping (1) defines a trajectory in $E^{(3)}$ which passes through $\mathbf{s}$

when $t = 0$.   On this trajectory, the velocity is

$$\frac{d\mathbf{x}}{dt} = \frac{\partial \mathbf{A}(\mathbf{s},t)}{\partial t} \tag{2}$$

and the acceleration is

$$\frac{d^2\mathbf{x}}{dt^2} = \frac{\partial^2}{\partial t^2} \mathbf{A}(\mathbf{s},t) \tag{3}$$

For $\mathbf{x}$ in C_t the mapping (1) has an inverse $\mathbf{A}^{-1}(\mathbf{x},t)$.   Let

$$\mathbf{V}(\mathbf{x},t) = \frac{\partial}{\partial t} \mathbf{A}(\mathbf{s},t) \Big]_{\mathbf{s} = \mathbf{A}^{-1}(\mathbf{x},t)}$$

Then

$$\frac{d\mathbf{x}}{dt} = \mathbf{V}(\mathbf{x},t) \tag{4}$$

The components of the acceleration can then be expressed as

$$\frac{d^2x_j}{dt^2} = \frac{\partial V_j}{\partial t} + \sum_{k=1}^{3} \frac{\partial V_j}{\partial x_k} V_k$$

or, in vector form,

$$\frac{d^2\mathbf{x}}{dt^2} = \frac{\partial \mathbf{V}}{\partial t} + (\mathbf{V} \cdot \boldsymbol{\nabla})\mathbf{V}$$

In general, if ξ is any quantity, scalar or vector, which is a function of $\mathbf{x}$ and of t, where $\mathbf{x}$ is given by (1), then

$$\frac{d\xi}{dt} = \frac{\partial \xi}{\partial t} + \sum_{k=1}^{3} V_k \frac{\partial \xi}{\partial x_k} = \frac{\partial \xi}{\partial t} + (\mathbf{V} \cdot \boldsymbol{\nabla})\xi$$

It is conventional, in these special circumstances, to abbreviate this expression by the symbol $D\xi/Dt$.

The jacobian $J(\mathbf{s},t)$ of (1) is the value of the determinant whose entry in the jth row and kth column is

$$J_{jk}(\mathbf{s},t) = \frac{\partial x_j}{\partial s_k} = \frac{\partial A_j(\mathbf{s},t)}{\partial s_k}$$

We shall assume it to be known that if $\mathbf{s}$ is in C_0, if R_0 is an open region in C_0 containing $\mathbf{s}$ and having the volume $m(R_0)$, if R_t is the image of R_0 in C_t having the volume $m(R_t)$, and if the diameter of R_0 is allowed to approach zero, then

$$\lim \frac{m(R_t)}{m(R_0)} = J(\mathbf{s},t)$$

More generally, if R_t is a region in C_t whose preimage in C_0 is R_0, and if $f(\mathbf{x})$ is a function defined on R_t, then

$$\int_{R_t} f(\mathbf{x}) \, dR_t = \int_{R_0} f(\mathbf{A}(\mathbf{s},t)) J(\mathbf{s},t) \, dR_0$$

where the integrals are volume integrals.

A fundamental relationship is that

$$\frac{1}{J}\frac{dJ}{dt} = \sum_{j=1}^{3} \frac{\partial V_j}{\partial x_j} = \nabla \cdot \mathbf{V} \tag{5}$$

To prove this, observe that

$$\frac{dJ_{jk}}{dt} = \frac{\partial V_j}{\partial s_k} = \sum_{i=1}^{3} \frac{\partial V_j J_{ik}}{\partial x_i}$$

But dJ/dt is the sum of three determinants, $J^{(j)}$, $j = 1, 2, 3$, obtained by replacing the elements J_{jk}, $k = 1, 2, 3$, in the jth row of J by derivatives dJ_{jk}/dt. Relationship (5) then follows from the elementary properties of determinants.

CONSERVATION OF MASS

Thus far no physical properties have been attributed to the continua C_t. Now we assign, for each C_t, a positive function $\rho(\mathbf{x},t)$, having the interpretation that if R_t is any open region in C_t, the total mass contained in R_t is the volume integral

$$\int_{R_t} \rho(\mathbf{x},t) \, dR_t$$

When $t = 0$, $\rho(\mathbf{x},0)$ will be written simply as $\rho(\mathbf{s})$.

The *conservation of mass* will be expressed by the statement that if R_0 is the preimage of R_t, then

$$\int_{R_t} \rho(\mathbf{x},t) \, dR_t = \int_{R_0} \rho(\mathbf{s}) \, dR_0$$

Since this equality holds for arbitrary R_t, we obtain

$$\rho(\mathbf{x},t) J(\mathbf{s},t) = \rho(\mathbf{s}) \qquad \text{for } \mathbf{x} = \mathbf{A}(\mathbf{s},t) \tag{6}$$

Equation (6) is the *equation of continuity*. A different form of the equation of continuity is found by differentiating (6) with respect to t; then

$$J \frac{D\rho}{Dt} + \rho \frac{dJ}{dt} = 0$$

or, from relationship (5),

$$\frac{D\rho}{Dt} + \rho\,\mathbf{\nabla}\cdot\mathbf{V} = 0 \tag{7}$$

When a density function and a family of continua have been defined in this fashion, we say that we are dealing with a *flow*.

Let the surface

$$G(x_1,x_2,x_3,t) = 0$$

be a connected component of the boundary of C_t. Taking the derivative with respect to time, we have $DG/Dt = 0$, or

$$\sum_{j=1}^{3} \frac{\partial G}{\partial x_j}\,V_j + \frac{\partial G}{\partial t} = 0$$

If $\hat{n}(\mathbf{x})$ is the unit normal vector to the boundary at $\mathbf{x}$, this condition becomes, in vector form,

$$\mathbf{V}\cdot\hat{n} + |\mathbf{\nabla}G|\frac{\partial G}{\partial t} = 0 \tag{8}$$

A large class of physical problems are described as *incompressible steady-state irrotational flows* with *fixed boundary*. "Incompressible" means that $D\rho/Dt = 0$. From the equation of continuity, this is equivalent to $\mathbf{\nabla}\cdot\mathbf{V} = 0$, which has the geometric interpretation that as the points of a region move along their trajectories, the volume of the changing region does not change. "Steady state" means that $\partial V/\partial t = 0$; in this case it is common to talk of the *streamlines*, which are the curves defined by the differential equations $dx_i/d\zeta = V_j(x)$, where ζ is an arbitrary parameter. "Irrotational" means that $\mathbf{\nabla}\times\mathbf{V} = 0$, which is equivalent to the existence of a *velocity potential* $\phi(\mathbf{x})$ such that $\mathbf{V} = -\mathbf{\nabla}\phi$. Finally, "fixed boundary" means that the connected components of the boundary, $G(\mathbf{x},t) = 0$, have $\partial G/\partial t = 0$. In this case the boundary condition (8) becomes $\mathbf{V}\cdot\hat{n} = 0$.

With these facts, the mathematical problem associated with this type of flow is to find a solution of Laplace's equation,

$$\nabla^2\phi = 0$$

subject to the condition on the normal derivative at the boundary

$$\frac{\partial\phi}{\partial n} = 0$$

There are basically two types of problems to which this representation is applicable. The first type are those problems in which the region containing the flow has points $\mathbf{x}$ for which $|\mathbf{x}|$ is arbitrarily large. In this case, the two conditions on ϕ are supplemented by specifying

$$\lim_{|\mathbf{x}| \to \infty} \nabla \phi$$

which is equivalent to specifying the velocity at very large distances.   In the second type a fixed boundary is arbitrarily added to the problem, and $\partial \phi / \partial n$ is specified on this boundary.   In this case the flow is studied on only one side of the boundary.   Typical problems of the first kind are those in which an obstacle is inserted in a constant flow ($\mathbf{V} = $ constant), and the resulting flow is determined under the assumption that a steady-state condition exists and the flow at points remote from the obstacle is not disturbed by it.   Typical problems of the second kind are those in which flow is created by the injection of material at a boundary surface.

All considerations that have been made up to this point have been a consequence of the mathematical formalism and the principle of conservation of mass.   We have seen that if certain hypotheses are made about the flow, it can be determined by the solution of certain partial-differential equations subject to certain boundary conditions.

DYNAMICS

If $\mathbf{x}$ is a point in C_t, the *force density* is defined to be a function $\mathbf{F}(\mathbf{x})$ such that the total force on an open region R_t in C_t is given by the volume integral

$$\int_{R_t} \mathbf{F}(\mathbf{x}) \, dR_t$$

The assumption is made that the total momentum contained in R_t is

$$\int_{R_t} \rho(\mathbf{x},t) \mathbf{V}(\mathbf{x},t) \, dR_t$$

and that the conservation of momentum can be applied in the form

$$\frac{d}{dt} \int_{R_t} \rho(\mathbf{x},t) \mathbf{V}(\mathbf{x},t) \, dR_t = \int_{R_t} \mathbf{F}(\mathbf{x}) \, dR_t$$

The left-hand member can be rewritten as

$$\frac{d}{dt} \int_{R_0} \rho(\mathbf{s}) \frac{\partial \mathbf{A}(\mathbf{s},t)}{\partial t} \, dR_0 = \int_{R_0} \rho(\mathbf{s}) \frac{\partial^2 A}{\partial t^2} \, dR_0 = \int_{R_t} \rho(\mathbf{x},t) \frac{d^2 \mathbf{x}}{dt^2} \, dR_t$$

Since R_t is arbitrary, the equation of motion becomes

$$\rho(\mathbf{x},t) \frac{d^2 \mathbf{x}}{dt^2} = \rho(\mathbf{x},t) \frac{D\mathbf{V}(\mathbf{x},t)}{Dt} = \mathbf{F}(\mathbf{x}) \tag{9}$$

The forces on R_t can be thought of as being the sum of *imposed* and *internal* forces.   The imposed forces are caused by sources external to the continuum and are vector fields depending on $\mathbf{x}$, $d\mathbf{x}/dt$, and t. The internal forces are caused by that part of C_t not in R_t exerting forces on R_t. It is assumed that these forces can be completely described in terms of stresses being exerted on the surface S_t of R_t. *Stresses* are forces per unit area, and the word "stress" is used instead of "pressure" to emphasize that in general these forces can have components tangential to S_t. It is further assumed that at any point $\mathbf{x}$ on S_t the stress depends only on $\mathbf{x}$ and on the normal $\hat{n}$ (in terms of microscopic physics, these assumptions are equivalent to the statement that the intermolecular forces have such a short range of influence that the forces at a point can be represented as the limit of the forces caused by molecules in an arbitrarily small neighborhood of the point). If $\sigma_i(\mathbf{x},\hat{n})$ is the ith component of the stress at $\mathbf{x}$ on the surface whose normal vector at $\mathbf{x}$ is $\hat{n}$, then there is a unique vector $\mathbf{\sigma}_i(\mathbf{x})$, with components σ_{ij}, for which

$$\sigma_i(\mathbf{x},\hat{n}) = \mathbf{\sigma}_i(\mathbf{x}) \cdot \hat{n} = \sum_{j=1}^{3} \sigma_{ij} n_j$$

The total force exerted on R_t by the stresses has the ith component

$$\int_{S_t} \mathbf{\sigma}_i \cdot \hat{n} \, dS_t$$

which can be transformed, by Green's theorem, to

$$\int_{R_t} \mathbf{\nabla} \cdot \mathbf{\sigma}_i \, dR_t = \int_{R_t} \sum_{j=1}^{3} \frac{\partial \sigma_{ij}}{\partial x_j} \, dR_t$$

Then, if there are no imposed forces, $\mathbf{F}(\mathbf{x})$ has the ith component

$$F_i(\mathbf{x}) = \mathbf{\nabla} \cdot \mathbf{\sigma}_i = \sum_{j=1}^{3} \frac{\partial \sigma_{ij}}{\partial x_j} \tag{10}$$

For any pair (i,j) the ijth moment of $\mathbf{F}(\mathbf{x})$ is

$$x_i F_j - x_j F_i = x_i \mathbf{\nabla} \cdot \mathbf{\sigma}_j - x_j \mathbf{\nabla} \cdot \mathbf{\sigma}_i = \mathbf{\nabla} \cdot (x_i \mathbf{\sigma}_j - x_j \mathbf{\sigma}_i) + \sigma_{ij} - \sigma_{ji}$$

Integration of this expression over R_t gives the total ijth moment on R_t,

$$\int_{S_t} (x_i \mathbf{\sigma}_j - x_j \mathbf{\sigma}_i) \cdot \hat{n} \, dS_t + \int_{R_t} (\sigma_{ij} - \sigma_{ji}) \, dR_t$$

The first integral is the moment produced by the stresses on S_t; thus the second integral is zero for arbitrary R_t, or

$$\sigma_{ij} = \sigma_{ji} \tag{11}$$

FLUID DYNAMICS

In an arbitrary continuum the quantity

$$p = -\frac{1}{3} \sum_{j=1}^{3} \sigma_{jj}$$

is called the *mean hydrostatic pressure*.   A *fluid* is characterized by the
assumption that

$$\sigma_{ij} = -p\delta_{ij} + \sigma'_{ij}$$

where σ'_{ij} depends only on the partial derivatives $\partial V_l/\partial x_m$ and is zero
when either $\mathbf{V} = $ constant or $\mathbf{V} = \mathbf{x} \times \mathbf{r}$, where $\mathbf{r}$ is a constant vector.
We refer to the case $\sigma'_{ij} = 0$ as the *hydrostatic case*.   The physical basis
for this characterization is that for a fluid in motion the deviation of the
stresses on a surface S_t from the hydrostatic case is caused by the *relative*
velocity between points on S_t and points very near S_t.

   In a great many situations it is useful to regard the term σ'_{ij} as
negligible and to take $\sigma_{ij} = -p\delta_{ij}$.   This is called the *hydrostatic approxi-*
mation.   In this approximation we write the equation of continuity (7)
and the equation of motion (9) [using (10)] in the forms

$$\frac{\partial \rho}{\partial t} + (\mathbf{V} \cdot \boldsymbol{\nabla})\rho + \rho \boldsymbol{\nabla} \cdot \mathbf{V} = 0 \tag{12}$$

$$\rho \left[\frac{\partial \mathbf{V}}{\partial t} + (\mathbf{V} \cdot \boldsymbol{\nabla})\mathbf{V} \right] + \boldsymbol{\nabla} p = 0 \tag{13}$$

If imposed forces are present, they must be added to the right-hand side of
(13).   Equations (12) and (13) are a system of partial-differential equations for the unknown functions ρ, p, and $\mathbf{V}$.   There are four scalar
equations for five unknown scalar functions, and therefore they cannot be
expected to determine the condition of the fluid.   In practice, this situation is remedied either by assuming a relationship between p and ρ or by
assuming a relationship among p, ρ, and the absolute temperature θ,
called the *equation of state*, and by adding a third scalar equation to (12)
and (13) from the laws of thermodynamics and the principle of conservation of energy.   In many practical situations it can be assumed that the
absolute temperature θ does not change during the fluid motion, or that
there is no exchange of heat between adjacent volumes of the fluid.   These
cases are referred to, respectively, as *isothermal* and *adiabatic flow* and are
among those cases in which p is a function of ρ.

   Even when these simplifying assumptions are made, the partial-
differential equations are not linear, and no general methods exist which
can be used to solve them.   When they must be solved, recourse is gen-

erally made to numerical-approximation techniques, made feasible by high-speed computer technology.

There is a special case which is of general interest. The fluid is first assumed to be uniform and at rest. That is, $\rho = \rho_0$, $p = p_0$, and $\mathbf{V} = 0$, where ρ_0 and p_0 are constant. A small disturbance is created in the fluid, so that $\rho = \rho_0 + \rho_1$, $p = p_0 + p_1$, and $\mathbf{V} = \mathbf{V}_1$, where ρ_1, p_1, and $\mathbf{V}_1$ are considered to be so small that their products and powers higher than the first can be disregarded. When these values are substituted into (12) and (13) and the higher powers are discarded, we obtain

$$\frac{\partial \rho_1}{\partial t} + \rho_0 \, \mathbf{\nabla} \cdot \mathbf{V}_1 = 0 \tag{12'}$$

$$\rho_0 \frac{\partial \mathbf{V}_1}{\partial t} + \mathbf{\nabla} p_1 = 0 \tag{13'}$$

$\mathbf{V}_1$ can be eliminated by applying the operators $\partial/\partial t$ and $\mathbf{\nabla} \cdot$ to (12') and (13'), respectively, and subtracting. Then

$$\frac{\partial^2 \rho_1}{\partial t^2} - \nabla^2 p_1 = 0$$

if p is a function of ρ,

$$p_1 = \frac{dp}{d\rho}\bigg]_{\rho = \rho_0} \rho_1$$

and so

$$\frac{\partial^2 \rho_1}{\partial t^2} - C^2 \, \nabla^2 \rho_1 = 0 \tag{14}$$

where

$$C^2 = \frac{dp}{d\rho}\bigg]_{\rho = \rho_0}$$

Equation (5) is the well-known *scalar wave equation*, used to describe the propagation of sound in a fluid. The pressure p_1 satisfies the same equation. Using the same approximation for p_1, we can eliminate ρ_1 from (12') and (13'), obtaining

$$\frac{\partial^2 \mathbf{V}_1}{\partial t^2} - C^2 \, \mathbf{\nabla}(\mathbf{\nabla} \cdot \mathbf{V}_1) = 0$$

This last equation is similar to the *vector wave equation*

$$\frac{\partial^2 \mathbf{V}}{\partial t^2} - C^2 \, \nabla^2 \mathbf{V} = 0 \qquad \nabla^2 \equiv \mathbf{\nabla}\mathbf{\nabla} \cdot \, - \mathbf{\nabla} \times \mathbf{\nabla} \times \tag{15}$$

From (13') we see that $\mathbf{\nabla} \times \mathbf{V}_1$ is identically zero if it is zero initially. In this case the equation for $\mathbf{V}_1$ is identical to (15). Solutions $\mathbf{V}$ for which $\mathbf{\nabla} \times \mathbf{V} = 0$ are called *longitudinal solutions*.

Returning to the description of the σ_{ij}, we next represent the σ'_{ij} as a linear combination of the $\partial V_l/\partial x_m$. This assumption is justified, for example, if the $\partial V_l/\partial x_m$ are so small that higher powers in an expansion for σ'_{ij} can be neglected. Let

$$\sigma'_{ij} = \sum_{\substack{l=1 \\ m=1}}^{3} a_{ijlm} \frac{\partial V_l}{\partial x_m}$$

The numbers a_{ijlm} may depend on $\mathbf{x}$, but they are independent of $\partial V_l/\partial x_m$. Since $\sigma'_{ij} = 0$ when $\mathbf{V}$ is of the form $\mathbf{x} \times \mathbf{r}$, where $\mathbf{r}$ is an arbitrary vector, $a_{ijlm} = a_{ijml}$, so that σ'_{ij} can be rewritten as

$$\sigma'_{ij} = \sum_{\substack{l=1 \\ m=1}}^{3} a_{ijlm} V_{lm}$$

where

$$V_{lm} = \frac{1}{2}\left(\frac{\partial V_l}{\partial x_m} + \frac{\partial V_m}{\partial x_l}\right)$$

Now, the total kinetic energy in an open region R_t is

$$T(R_t) = \int_{R_t} \frac{1}{2}\rho(\mathbf{x},t)|\mathbf{V}(\mathbf{x},t)|^2 \, dR_t$$

Therefore

$$\frac{d}{dt} T(R_t) = \frac{d}{dt} \int_{R_0} \frac{1}{2}\rho(\mathbf{s}) \left|\frac{\partial \mathbf{A}(\mathbf{s},t)}{\partial t}\right|^2 dR_0$$

$$- \int_{R_0} \rho(\mathbf{s}) \frac{\partial^2 \mathbf{A}}{\partial t^2} \cdot \frac{\partial \mathbf{A}}{\partial t} \, dR_0 - \int_{R_t} \mathbf{F}(\mathbf{x}) \cdot \mathbf{V}(\mathbf{x},t) \, dR_t$$

$$= \int_{R_t} \sum_{i=1}^{3} V_i \mathbf{\nabla} \cdot \boldsymbol{\delta}_i \, dR_t$$

Using Green's theorem, we have

$$\frac{d}{dt} T(R_t) = \int_{S_t} \sum_{i=1}^{3} V_i \boldsymbol{\delta}_i \cdot \hat{n} \, dS_t - \int_{R_t} \sum_{i=1}^{3} \boldsymbol{\delta}_i \cdot \mathbf{\nabla} V_i \, dR_t$$

The first integral is the rate at which work is being performed on R_t by external forces and can be symbolized as $dW(R_t)/dt$. The second integral can be written as

$$- \int_{R_t} \sum_{\substack{i=1 \\ j=1}}^{3} \sigma_{ij} \frac{\partial V_i}{\partial x_j} \, dR_t$$

and because $\sigma_{ij} = \sigma_{ji}$ [from (11)], this can be rewritten as

$$- \int_{R_t} \sum_{\substack{i=1 \\ j=1}}^{3} \sigma_{ij} V_{ij} \, dR_t = \int_{R_t} p \sum_{j=1}^{3} \frac{\partial V_j}{\partial x_j} \, dR_t - \int_{R_t} \sum_{\substack{i=1 \\ j=1}}^{3} \sum_{\substack{l=1 \\ m=1}}^{3} a_{ijlm} V_{lm} V_{ij}$$

Recall the relationship

$$\nabla \cdot \mathbf{V} = \frac{1}{J} \frac{dJ}{dt}$$

where J is the jacobian; the first integral on the right-hand side of the above equality is the rate at which the potential energy in R_t is changing. We write this as $-dU(R_t)/dt$, since the potential energy decreases under expansion. Then

$$\frac{dW(R_t)}{dt} + \int_{R_t} \sum_{\substack{i=1 \\ j=1}}^{3} \sum_{\substack{l=1 \\ m=1}}^{3} a_{ijlm} V_{lm} V_{ij} \, dR_t = \frac{dT(R_t)}{dt} + \frac{dU(R_t)}{dt}$$

Thus the integral must represent the rate at which heat is being created in the region R_t. The physical explanation is that this heat is a combination of the heat produced by frictional forces and the heat flowing into (or out of) R_t from the adjacent region. The value of the a_{ijlm} may therefore depend upon the nature of the thermodynamic processes, so that, for example, the values under adiabatic conditions may be different from those under isothermal conditions. In any event, the quantity

$$\Psi = \sum_{\substack{i=1 \\ j=1}}^{3} a_{ijlm} V_{lm} V_{ij}$$

called the *dissipation function*, represents the heat that is being created per unit volume, and

$$\sigma'_{ij} = \frac{\partial \Psi}{\partial V_{ij}}$$

Ψ is a quadratic form in the V_{lm}. Its value should be unchanged when the coordinate axes are rotated. We can verify by linear algebra that this implies that Ψ can be written as

$$\Psi = \frac{\lambda}{2} \left(\sum_{j=1}^{3} V_{jj} \right)^2 + \mu \sum_{\substack{i=1 \\ j=1}}^{3} V_{ij}^2$$

where λ and μ are independent of the V_{ij}. Then

$$\sigma_{ij} = -p\delta_{ij} + 2\mu V_{ij} + \lambda \delta_{ij} \sum_{k=1}^{3} V_{kk}$$

Recalling that

$$p = -\frac{1}{3} \sum_{j=1}^{3} \sigma_{jj}$$

we have

$$3\lambda + 2\mu = 0$$

and so, finally,

$$\sigma_{ij} = -p\delta_{ij} + \mu \left(\frac{\partial V_i}{\partial x_j} + \frac{\partial V_j}{\partial x_i}\right) - \frac{2\mu}{3} \delta_{ij} \sum_{k=1}^{3} \frac{\partial V_k}{\partial x_k}$$

The quantity μ is called the *viscosity* of the fluid.

In general, μ can vary in the fluid. As a useful approximation, μ is often taken to be a constant. In this case

$$F_i(\mathbf{x}) = -\frac{\partial p}{\partial x_i} + \mu \sum_{j=1}^{3} \frac{\partial^2 V_i}{\partial x_j^2} + \frac{1}{3} \mu \frac{\partial}{\partial x_i} \sum_{j=1}^{3} \frac{\partial V_j}{\partial x_j}$$

The vector form of the equation of motion is then

$$\rho(\mathbf{x},t) \left[\frac{\partial \mathbf{V}}{\partial t} + (\mathbf{V} \cdot \nabla)\mathbf{V}\right] + \nabla p = \mu \left[\nabla^2 \mathbf{V} + \tfrac{1}{3} \nabla (\nabla \cdot \mathbf{V})\right] \qquad (16)$$

This is the *Stokes-Navier equation*. The hydrostatic approximation is equivalent to assuming that $\mu = 0$. As in the case of the hydrostatic approximation, further assumptions must be made about the thermodynamic processes before (16) can be used, with the equation of continuity, to determine the flow. However, under various assumptions, (16) can be used to generate linear partial-differential equations to which general methods of solution can be applied.

ELASTICITY

The methods and formalism in the theory of elasticity are similar to those in fluid dynamics. However, whereas in fluid dynamics the relative *velocity* of nearby points is the basis for the description of the stresses, in the theory of elasticity the stresses are caused by the *position* of points. The elastic medium is said to be in an *unstrained state* if the stresses are zero. We shall assume that the continuum C_0 represents the unstrained state of the medium. An elastic medium is characterized by the fact that when it is not in an unstrained state, the stresses act to restore the medium to an unstrained state.

Let us rewrite the mapping (1) in the form

$$\mathbf{x} = \mathbf{s} + \mathbf{B}(\mathbf{s},t) \tag{17}$$

The vector $\mathbf{B}(\mathbf{s},t)$ is called the *displacement* of $\mathbf{s}$. When $\mathbf{B}(\mathbf{s},t) \neq 0$, the continuum C_t is said to represent the medium in a *state of strain*.

Let us consider the one-dimensional case; that is, let the mapping (8) be of the form

$$x_1 = s_1 + B(s_1,t) \qquad x_2 = s_2 \qquad x_3 = s_3 \tag{18}$$

The restoring force will have only the component $F_1(x_1)$, which, by the general theory, must be of the form $F_1 = \partial\sigma/\partial x_1$. Any interval I_0 on the s_1 axis will be transformed by (17) into an interval I_t, and if $l(I_0)$ and $l(I_t)$ are the lengths of these intervals, the limit is

$$\lim_{l(I_0)\to 0} \frac{l(I_t) - l(I_0)}{l(I_0)} = \frac{dx_1}{ds_1} - 1 = \frac{\partial B}{\partial s_1}$$

It is this *relative distortion of length*, called the *strain*, which, by assumption, determines the stress. If the strained state (18) is maintained by the application of external forces, the medium will be in a state of thermodynamic equilibrium. If we refer all thermodynamic functions to the unstrained state, the potential energy per unit volume will depend on $\partial B/\partial s_1$ (along with other thermodynamic variables) and will be zero when $\partial B/\partial s_1 = 0$.

Let I_t be an interval on the x_1 axis, whose preimage is I_0, and let R_t be the region in C_t consisting of a fixed area A in the $x_2 x_3$ plane (identical to the $s_2 s_3$ plane), which moves along the interval I_t. Let R_0 be the preimage of R_t. The kinetic energy in R_t is

$$T(I_t) = \int_{I_t} \frac{1}{2}\, \rho(\mathbf{x},t) \left(\frac{dx_1}{dt}\right)^2 dx_1\, A = \int_{I_0} \frac{1}{2}\, \rho(\mathbf{s}) \left[\frac{\partial B(s_1,t)}{\partial t}\right]^2 ds_1\, A$$

and

$$\begin{aligned}
\frac{dT(I_t)}{dt} &= \int_{I_0} \rho(\mathbf{s})\, \frac{\partial^2 B}{\partial t^2} \frac{\partial B}{dt}\, ds_1\, A = \int_{I_t} F_1(x_1)\, \frac{\partial B}{dt}\, dx_1\, A \\
&= \int_{I_t} \frac{\partial\sigma}{\partial x_1} \frac{\partial B}{\partial t}\, dx_1\, A \\
&= \left.\sigma \frac{\partial B}{dt}\right]_{I_t} - \int_{I_t} \sigma \frac{\partial}{\partial x_1}\left(\frac{\partial B}{\partial t}\right) dx_1\, A \\
&= \left.\sigma \frac{\partial B}{\partial t}\right]_{I_t} - \int_{I_0} \sigma \frac{\partial}{\partial t}\left(\frac{\partial B}{\partial s_1}\right) ds_1\, A
\end{aligned}$$

In the last of these expressions, the first term represents $dW\,(I_t)/dt$, the rate at which work is being done on I_t by external forces, and the second term is an integral on the unstrained continuum C_0. If dt is a small change in time, then

$$dT = dW - \int_{I_0} \sigma \, d\left(\frac{\partial B}{\partial s_1}\right) ds_1 \, A$$

Then if dT^*, dW^*, dU^*, and dQ^* are, respectively, the changes in kinetic energy, work done by external forces, potential energy, and heat *per unit volume of the unstrained medium*, we have

$$dT^* = dW^* - \sigma \, d\left(\frac{\partial B}{\partial s_1}\right)$$

From the first law of thermodynamics,

$$dW^* + dQ^* = dT^* + dU^*$$

and so

$$\sigma \, d\left(\frac{\partial B}{\partial s_1}\right) = dU^* - dQ^*$$

Now, dU^* is the differential of a function of the thermodynamic variables, which include $\partial B/\partial s_1$. If the change in the time, dt, is adiabatic, then $dQ^* = 0$. If the change is isothermal, then, from the second law of thermodynamics, dQ^* is the differential of a function of the thermodynamic variables. In either case, there exists a function Φ, the *strain-energy function*, depending on the thermodynamic variables, such that

$$\sigma = \frac{\partial \Phi}{\partial(\partial B/\partial s_1)}$$

Φ must be a minimum when $\partial B/\partial s_1 = 0$, and σ must be zero when $\partial B/\partial s_1 = 0$. Therefore Φ, when expanded in powers of $\partial B/\partial s_1$, must have the form

$$\Phi = \Phi_0 + \frac{E}{2}\left(\frac{\partial B}{\partial s_1}\right)^2 + \cdots$$

where E is independent of $\partial B/\partial s_1$. If the unstrained medium is homogeneous, E will be a constant, depending only on whether the thermodynamic processes are adiabatic or isothermal. If higher powers than those listed are disregarded, then

$$\sigma = E \frac{\partial B}{\partial s_1}$$

which is the familiar *Hooke's law;* that is, the stress is proportional to the strain. The equation of motion becomes

$$\frac{\partial^2 B}{\partial t^2} = E \frac{\partial^2 B}{\partial s_1{}^2}$$

the familiar wave equation, which in this case describes the propagation of longitudinal disturbances in a homogeneous elastic rod. If the medium is maintained in equilibrium by the application of external forces, then $\partial^2 B/\partial s_1{}^2 = 0$, which limits the form of the displacement to $B = a + bs_1$, implying that σ is constant.

Returning to the general case, from the mapping (1), we write the components of the differential of $\mathbf{x}$ as

$$dx_j = ds_j + \sum_{k=1}^{3} \frac{\partial B_j}{\partial s_k} ds_k \qquad \text{for } j = 1, 2, 3$$

Then

$$\sum_{j=1}^{3} (dx_j)^2 = \sum_{j=1}^{3} (ds_j)^2 + 2 \sum_{\substack{i=1 \\ j=1}}^{3} B_{ij}\, ds_i\, ds_j$$

where

$$B_{ij} = \frac{1}{2} \left(\frac{\partial B_i}{\partial s_j} + \frac{\partial B_j}{\partial s_i} + \sum_{k=1}^{3} \frac{\partial B_k}{\partial s_i} \frac{\partial B_k}{\partial s_j} \right)$$

From this it follows that if Γ_0 is a line segment in C_0, having length $l(\Gamma_0)$ and direction cosines $\cos \alpha_j$, $j = 1, 2, 3$, and if Γ_t is the image of Γ_0 in C_t, having length $l(\Gamma_t)$, then

$$\lim_{l(\Gamma_0) \to 0} \frac{l(\Gamma_t) - l(\Gamma_0)}{l(\Gamma_0)} = \left(1 + 2 \sum_{\substack{i=1 \\ j=1}}^{3} B_{ij} \cos \alpha_i \cos \alpha_j \right)^{\frac{1}{2}} - 1$$

The B_{ij}, together with the direction of Γ_0, completely determine this limit and are therefore assumed to determine the σ_{ij}. The B_{ij} are called the *coefficients of strain*.

At this point it is further assumed that the derivatives $\partial B_i/\partial s_j$ are so small that in any expression containing powers of $\partial B_i/\partial s_j$ only the lowest powers need be retained. In particular, the above expression for B_{ij} is replaced by

$$B_{ij} = \frac{1}{2} \left(\frac{\partial B_i}{\partial s_j} + \frac{\partial B_i}{\partial s_i} \right)$$

By means of computations similar to those carried out in the theory of fluid mechanics and arguments similar to those employed in the one-dimensional case, the existence of a strain-energy function Φ is then established for adiabatic or isothermal processes, and

$$\sigma_{ij} = \frac{\partial \Phi}{\partial B_{ij}}$$

If we regard Φ as a quadratic form in the B_{ij}, for an isotropic medium (a crystalline material, for example, is not isotropic) Φ must remain invariant under rotation of the coordinate system, and we may write

$$\Phi = \frac{\lambda_1}{2} \left(\sum_{j=1}^{3} B_{jj} \right)^2 + \lambda_2 \sum_{\substack{i=1 \\ j=1}} B_{ij}{}^2$$

where λ_1 and λ_2 will be different in the adiabatic and isothermal cases. In either event,

$$\sigma_{ij} = \lambda_1 \delta_{ij} \sum_{k=1}^{3} B_{kk} + 2\lambda_2 B_{ij}$$

which is Hooke's law for an isotropic medium. If we define, as in the case of fluids,

$$p = -\tfrac{1}{3} \sum_{j=1}^{3} \sigma_{jj}$$

we have

$$p = -(\lambda_1 + \tfrac{2}{3}\lambda_2) \sum_{k=1}^{3} B_{kk}$$

Then

$$\sigma_{ij} = -\delta_{ij} p$$

is equivalent to

$$2\lambda_2 \left(B_{ij} - \tfrac{1}{3}\delta_{ij} \sum_{k=1}^{3} B_{kk} \right) = 0$$

Thus a hydrostatic fluid can be regarded as an elastic medium for which the constant $\lambda_2 = 0$.

From the above equation for σ_{ij}, the equation of motion becomes, in vector form,

$$\rho(s) \frac{\partial^2 \mathbf{B}}{\partial t^2} = (\lambda_1 + \lambda_2) \, \mathbf{\nabla} (\mathbf{\nabla} \cdot \mathbf{B}) + \lambda_2 \, \nabla^2 \mathbf{B} \tag{19}$$

If the medium is maintained in equilibrium by the application of external forces, then

$$(\lambda_1 + \lambda_2) \, \mathbf{\nabla} (\mathbf{\nabla} \cdot \mathbf{B}) + \lambda_2 \, \nabla^2 \mathbf{B} = 0 \tag{20}$$

is the linear partial-differential equation that must be satisfied by $\mathbf{B}$. Thus if a known set of forces is applied to the boundary of C_0, equation

(20), together with the boundary conditions which equate the sum

$$\sum_{j=1}^{3} \sigma_{ij} n_j$$

to the ith component of the externally applied stress, determines the equilibrium configuration.   This is the fundamental problem in elasticity. If $\rho(s)$ is a constant, (19) is a general form of the vector wave equation. In particular, for *longitudinal* solutions, $\boldsymbol{\nabla} \times \mathbf{B} = 0$, the velocity of propagation C_L satisfies

$$C_L{}^2 = \frac{\lambda_1 + 2\lambda_2}{\rho}$$

and for *transverse* solutions, $\boldsymbol{\nabla} \cdot \mathbf{B} = 0$, the velocity of propagation $C_T{}^2$, satisfies

$$C_T{}^2 = \frac{\lambda_2}{\rho}$$

Electromagnetic Theory

THE ELECTROMAGNETIC FIELD

The partial-differential equations of electromagnetic theory are not derived, but are given axiomatically. In a connected open region R of three-dimensional euclidean space, whose points are designated by vectors $\mathbf{x}$, and for t a real variable on some interval, these equations are

$$\nabla \times \mathbf{E}(\mathbf{x},t) + \frac{\partial \mathbf{B}(\mathbf{x},t)}{\partial t} = 0$$

$$\nabla \times \mathbf{H}(\mathbf{x},t) - \frac{\partial \mathbf{D}}{\partial t}(\mathbf{x},t) = \mathbf{J}(\mathbf{x},t)$$

$$\nabla \cdot \mathbf{B} = 0 \tag{1}$$

$$\nabla \cdot \mathbf{D} = \rho(\mathbf{x},t)$$

$$\nabla \cdot \mathbf{J} + \frac{\partial \rho}{\partial t} = 0$$

In these equations $\mathbf{E}$ and $\mathbf{H}$ are called the *electric* and *magnetic field vectors;* $\mathbf{D}$ and $\mathbf{B}$ are called the *electric* and *magnetic flux densities;* $\mathbf{J}$ is called the *current density;* and ρ is called the *charge density.* The functions and the derivatives which appear in (1) are assumed to be continuous and bounded in R. As points in R approach points $\mathbf{x}$ on the boundary of R, the functions in (1) have limiting values which we shall call the *values of the functions at* $\mathbf{x}$.

If R_1 and R_2 are disjoint, connected open regions having a surface S as a common boundary, if S has a continuous normal vector $\hat{n}$, and if $\mathbf{E}_1$ and $\mathbf{H}_1$ and $\mathbf{E}_2$ and $\mathbf{H}_2$ are the electric and magnetic field vectors in R_1 and R_2, respectively, it is further assumed that on S

$$\hat{n} \times \mathbf{E}_1 = \hat{n} \times \mathbf{E}_2 \qquad \hat{n} \times \mathbf{H}_1 = \hat{n} \times \mathbf{H}_2 \tag{2}$$

The pair of vector fields $(\mathbf{E},\mathbf{H})$ is called the *electromagnetic field.*

The physical content of electromagnetic theory is given by correct interpretation of the quantities in (1). First, it should be recognized that electromagnetic fields are produced by charged particles, either at rest or in motion. Second, electromagnetic fields define forces which act on charged particles. Electromagnetic fields are measured by performing measurements of the position and motion of charged particles. The status of equations (1) and conditions (2) as a physical theory rests upon their ability to predict and explain these measurements. They play the same epistemological role in the study of electromagnetic forces as do Newton's laws in the study of mechanical forces.

The charge density ρ is given the meaning that in an arbitrary region R' the total electric charge is the volume integral

$$\int_{R'} \rho \, dR'(\mathbf{x})$$

The current density $\mathbf{J}$ is given the meaning that for an arbitrary surface S', having a continuous unit normal vector $\hat{n}$, the rate at which charge is crossing S' in the direction $\hat{n}$ is the surface integral

$$\int_{S'} \mathbf{J} \cdot \hat{n} \, dS'$$

The last equation of (1) is an expression of the fact that electric charge is conserved and, in analogy with continuum dynamics, it is called the *equation of continuity.*

The matter which occupies the region R will have electromagnetic properties that determine the relationship of $\mathbf{D}$ and $\mathbf{B}$ to $\mathbf{E}$ and $\mathbf{H}$, respectively. In *free space* this relationship is

$$\mathbf{D} = \epsilon_0 \mathbf{E} \qquad \mathbf{B} = \mu_0 \mathbf{H}$$

where ϵ_0 and μ_0 are real constants. In the most general case, $\mathbf{D}(\mathbf{x},t)$ and $\mathbf{B}(\mathbf{x},t)$ depend, respectively, on the values of $\mathbf{E}(\mathbf{x},t')$ and $\mathbf{H}(\mathbf{x},t')$ for all $t' \leq t$. We shall limit our considerations to those cases in which $\mathbf{D} = \epsilon \mathbf{E}$

and $\mathbf{B} = \mu\mathbf{H}$, where ϵ and μ are real constants.   In this case R is said to be *isotropic* and *homogeneous*.

The connection between the electromagnetic field and the forces on charged particles is given by the assumption that for an arbitrary region R', contained in a homogeneous and isotropic region R, the forces on R' are given by the volume integral

$$\int_{R'} (\rho\mathbf{E} + \mu\mathbf{J} \times \mathbf{H}) \, dR'$$

When the first two equations of (1), which are called *Maxwell's equations*, are put into integral form by means of Stokes' theorem, they represent the laws of Faraday and Herz; the assumption of conditions (2) is equivalent to the assumption that these laws are valid in the presence of discontinuities in the electromagnetic properties of matter.   The third and fourth equations of (1), when converted into integral form by means of Green's theorem, express the fact that the total magnetic flux leaving a volume is zero and the total electric flux leaving a volume is the total charge contained in the volume.

The problems of electromagnetic theory are to find solutions of (1) which satisfy certain boundary conditions.   We shall show here the mathematical and physical reasoning which leads to the formulation of meaningful boundary-value problems.

THE HARMONIC CASE

We shall further restrict our considerations to the case in which the time dependence of the quantities which appear in (1) is a linear combination of $\cos \omega t$ and $\sin \omega t$, where the coefficients are functions of $\mathbf{x}$.   The quantity $\omega/2\pi$ is called the *oscillation frequency*.   This case is called the *harmonic*, or *monochromatic*, *case*.   The positive number $k = \omega \sqrt{\epsilon\mu}$ is called the *propagation constant*.

The first two equations in (1) become

$$\begin{aligned} \boldsymbol{\nabla} \times \mathbf{E}(\mathbf{x}) - i\mu\omega\mathbf{H}(\mathbf{x}) &= 0 \\ \boldsymbol{\nabla} \times \mathbf{H}(\mathbf{x}) + i\epsilon\omega\mathbf{E}(\mathbf{x}) &= \mathbf{J}(\mathbf{x}) \end{aligned} \tag{3}$$

where the vectors are *complex vectors* and are related to the vectors in (1) by, for example, the equation

$$\mathbf{E}(\mathbf{x},t) = \tfrac{1}{2}[\mathbf{E}(\mathbf{x})e^{-i\omega t} + \tilde{\mathbf{E}}(\mathbf{x})e^{i\omega t}]$$

where $\tilde{\mathbf{E}}(\mathbf{x})$ is the complex conjugate of $\mathbf{E}(\mathbf{x})$.   The equation of continuity becomes

$$\boldsymbol{\nabla} \cdot \mathbf{J} - i\omega\rho = 0$$

The remaining equations in (1), $\nabla \cdot \mathbf{H} = 0$ and $\nabla \cdot \mathbf{E} = \rho/\epsilon$, are then derivable from (3) and the equation of continuity; therefore, in this case, they are not assumptions.   Conditions (2) continue to hold.

Another way to obtain (2) is to regard, for example, $\mathbf{E}(\mathbf{x})$ as being the Fourier transform of $\mathbf{E}(\mathbf{x},t)$,

$$\int_{-\infty}^{\infty} \mathbf{E}(\mathbf{x},t) e^{i\omega t}\, dt$$

Then (2) is obtained by formally taking the Fourier transform of (1), without regard to the existence of the integrals.   If care is taken to establish the existence of the various Fourier integrals, the solution of the general time-dependent problem can be found by solving (3) for all ω.   In this case, however, it should be borne in mind that in most materials different from free space, even those regarded as homogeneous and isotropic, the relationships $\mathbf{D} = \epsilon\mathbf{E}$ and $\mathbf{B} = \mu\mathbf{H}$ are only approximate, with the result that the μ and the ϵ which appear in (3) are dependent on ω.   In practice, this dependence can be neglected for very wide ranges of ω, but in performing the Fourier inversion it may be a critical factor.

RADIATION

We shall assume that $\mathbf{J}$ in equation (3) is zero, except on a finite subregion of R.   If $\mathbf{x}$ is in R, then the vector field

$$\mathbf{A}(\mathbf{x}) = -\int_R \mathbf{J}(\mathbf{x}') \phi(\mathbf{x},\mathbf{x}')\, dR(\mathbf{x}')$$

where

$$\phi(\mathbf{x},\mathbf{x}') = \frac{1}{4\pi} \frac{e^{ik|\mathbf{x}-\mathbf{x}'|}}{|\mathbf{x} - \mathbf{x}'|}$$

is called the *vector potential* generated by $\mathbf{J}$.   We shall show that the vector fields

$$\mathbf{H}(\mathbf{x}) = \nabla \times \mathbf{A}(\mathbf{x})$$
$$\mathbf{E}(\mathbf{x}) = -\frac{1}{i\omega\epsilon} [\nabla(\nabla \cdot \mathbf{A}) + k^2\mathbf{A}] \tag{4}$$

are solutions of (3).   The proof depends only upon showing that

$$\nabla \times \nabla \times \mathbf{A} - \nabla(\nabla \cdot \mathbf{A}) - k^2\mathbf{A} = \mathbf{J}$$

or that for each of the rectangular components A_j of $\mathbf{A}$ and J_j of $\mathbf{J}$,

$$\nabla^2 A_j + k^2 A_j = -J_j$$

Thus the problem is to show that if f is a continuous function in R and $f = 0$ except on a finite subregion, the function

$$\psi(\mathbf{x}) = \int_R f(\mathbf{x}')\phi(\mathbf{x},\mathbf{x}') \, dR(\mathbf{x}')$$

satisfies the equation

$$\nabla^2\psi + k^2\psi = f$$

First, observe that for $\mathbf{x} \neq \mathbf{x}'$

$$\nabla^2\phi + k^2\phi = 0$$

Second, observe that although $\phi(\mathbf{x},\mathbf{x}')$ and $\boldsymbol{\nabla}\phi(\mathbf{x},\mathbf{x}')$ have singularities at $\mathbf{x} = \mathbf{x}'$, the volume integral which defines $\psi(\mathbf{x})$ is a proper integral, as in the volume integral for $\boldsymbol{\nabla}\psi$; that is,

$$\boldsymbol{\nabla}\psi(\mathbf{x}) = \int_R f(\mathbf{x}') \, \boldsymbol{\nabla}\phi(\mathbf{x},\mathbf{x}') \, dR(\mathbf{x}')$$

Let $\mathbf{x}_0$ be in R; let R_δ be a sphere in R with center at $\mathbf{x}_0$ and radius δ; let S_δ be the surface of R_δ, with unit outward normal $\hat{n}$. Then

$$\int_{S_\delta} \boldsymbol{\nabla}\psi(\mathbf{x}) \cdot \hat{n}(\mathbf{x}) \, dS_\delta(\mathbf{x}) = \int_{R_\delta} f(\mathbf{x}') \int_{S_\delta} \boldsymbol{\nabla}\phi(\mathbf{x},\mathbf{x}') \cdot \hat{n}(\mathbf{x}) \, dS_\delta(\mathbf{x}) \, dR(\mathbf{x}')$$
$$+ \int_{CR_\delta} f(\mathbf{x}') \int_{S_\delta} \boldsymbol{\nabla}\phi(\mathbf{x},\mathbf{x}') \cdot \hat{n}(\mathbf{x}) \, dS_\delta(\mathbf{x}) \, dR(\mathbf{x}')$$

where CR_δ is that part of R which is exterior to R_δ. In the integral over CR_δ the inner integral can be transformed by Green's theorem. In the integral over R_δ the inner integral can be evaluated directly to be $1 + g(\delta)$, where

$$\lim_{\delta \to 0} g(\delta) = 0$$

Thus

$$\int_{S_\delta} \boldsymbol{\nabla}\psi(\mathbf{x}) \cdot \hat{n}(\mathbf{x}) \, dS_\delta(\mathbf{x}) = \int_{R_\delta} \left\{ f(\mathbf{x})[1 + g(\delta)] - k^2[\psi(\mathbf{x})] \right.$$
$$\left. - \int_{R_\delta} f(\mathbf{x}')\phi(\mathbf{x},\mathbf{x}') \, dR(\mathbf{x}') \right\} dR(\mathbf{x})$$

If V_δ is the volume of R_δ, then

$$\nabla^2\psi(\mathbf{x}_0) = \lim_{\delta \to \infty} \frac{1}{V_\delta} \int_{S_\delta} \boldsymbol{\nabla}\psi(\mathbf{x}) \cdot \hat{n}(\mathbf{x}) \, dS_\delta(\mathbf{x}) = f(\mathbf{x}_0) - k^2\psi(\mathbf{x}_0)$$

This completes the proof.

There may, of course, be other solutions of (3) in R. However, if R is the entire space, the solutions (4) have a special property which distinguishes them from the other solutions. This property centers about the form of the field when $|\mathbf{x}|$ is very large. Let $\mathbf{x} = r\hat{u}$, where $\hat{u}$ is an arbitrary unit vector. If $r > \sqrt{2}\,|\mathbf{x}'|$ for all $\mathbf{x}'$ at which $\mathbf{J}(\mathbf{x}')$ does not vanish, then $|\mathbf{x} - \mathbf{x}'|$ can be written as

$$|\mathbf{x} - \mathbf{x}'| = r - \hat{u} \cdot \mathbf{x}' + \cdots$$

a series in descending powers of r whose convergence is uniform in $\hat{u}$. Then, from (4),

$$\mathbf{H}(\mathbf{x}) = \frac{ike^{ikr}}{4\pi r} \int_R \exp(-ik\hat{u} \cdot \mathbf{x}')\,\mathbf{J}(\mathbf{x}')(1 + \cdots)\,dR(\mathbf{x}') \times \hat{u}$$

$$\mathbf{E}(\mathbf{x}) = \sqrt{\frac{\mu}{\epsilon}} \left\{ \frac{ike^{ikr}}{4\pi r} \int_R \exp(ik\hat{u} \cdot \mathbf{x}')\,[\mathbf{J}(\mathbf{x}') + \cdots]\,dR(\mathbf{x}') \times \hat{u} \right\} \times \hat{u}$$

$$(5)$$

where the indicated series are in descending powers of r and their convergence is uniform in $\hat{u}$. The following conditions are satisfied by $\mathbf{E}$ and $\mathbf{H}$:

$r|\mathbf{E}|$ and $r|\mathbf{H}|$ have a bound which is independent of $\hat{u}$

$$r\left(\mathbf{E} + \sqrt{\frac{\mu}{\epsilon}}\,\hat{u} \times \mathbf{H} \right) \to 0 \text{ uniformly in } \hat{u}$$

$$r\left(\hat{u} \times \mathbf{E} - \sqrt{\frac{\mu}{\epsilon}}\,\mathbf{H} \right) \to 0 \text{ uniformly in } \hat{u}$$

as $r \to \infty$

$$(6)$$

These conditions are called the *Sommerfeld radiation conditions.*

We shall now see that of all the solutions of (3) in R, only the solutions given by (4) satisfy the radiation conditions. This is equivalent to showing that if $\mathbf{J} = 0$ in R, the only solutions to (3) which satisfy (5) are $\mathbf{E} = \mathbf{H} = 0$. Let $\mathbf{x}_0$ be in R, and let S_δ be the surface of a sphere with center at $\mathbf{x}_0$ and radius δ. Let $\hat{n}$ be the unit normal, directed into R, on S_δ. In that part of R exterior to S_δ, let

$$\mathbf{C} = \phi(\mathbf{x},\mathbf{x}_0)\mathbf{a}$$

where $\mathbf{a}$ is an arbitrary constant vector. Then

$$\nabla \times \mathbf{C} = \nabla\phi \times \mathbf{a} \qquad \nabla \times \nabla \times \mathbf{C} - k^2\mathbf{C} = \nabla(\mathbf{a} \cdot \nabla\phi)$$

If $\mathbf{J} = 0$, then, from (3),

$$\nabla \times \nabla \times \mathbf{E} - k^2\mathbf{E} = 0 \qquad \nabla \cdot \mathbf{E} = 0$$

Thus

$$\nabla \cdot [\phi\mathbf{a} \times (\nabla \times \mathbf{E}) - \mathbf{E} \times (\nabla\phi \times \mathbf{a}) - \mathbf{E}(\mathbf{a} \cdot \nabla\phi)]$$
$$= \mathbf{E} \cdot (\nabla \times \nabla \times \mathbf{C}) - \mathbf{C} \cdot (\nabla \times \nabla \times \mathbf{E}) - \mathbf{E} \cdot \nabla(\mathbf{a} \cdot \nabla\phi) = 0$$

Let S_r be the surface of the sphere with center at $\mathbf{x} = 0$, with outward unit normal $\hat{u}$ and with radius $r > |\mathbf{x}_0| + \delta$. By Green's theorem,

$$\mathbf{a} \cdot \left\{ \int_{S_r} [\phi(\nabla \times E) \times \hat{u} - (\hat{u} \times \mathbf{E}) \times \nabla\phi - \hat{u} \cdot \mathbf{E}\,\nabla\phi]\,dS_r \right\}$$

$$= \mathbf{a} \cdot \left\{ \int_{S_\delta} [\phi(\nabla \times E) \times \hat{u} - (\hat{u} \times \mathbf{E}) \times \nabla\phi - (\hat{u} \cdot \mathbf{E})\,\nabla\phi]\,dS_\delta \right\}$$

On S_δ,

$$\phi = \frac{1}{\delta} + \cdots \qquad \nabla\phi = -\frac{1}{\delta^2}\hat{n} + \cdots$$

and both series are in ascending powers of δ and are uniform in $\hat{n}$. Consequently, the integral over S_δ converges to $\mathbf{E}(\mathbf{x}_0)$ as $\delta \to 0$. Since $\mathbf{a}$ is arbitrary,

$$\mathbf{E}(\mathbf{x}_0) = \int_{S_r} [i\omega\mu(\mathbf{H} \times \hat{u})\phi - (\hat{u} \times \mathbf{E}) \times \nabla\phi - (\hat{u} \cdot \mathbf{E})\,\nabla\phi]\,dS_r(\mathbf{x})$$

On S_r,

$$\phi = \frac{e^{ikr}}{4\pi r}\exp(-ik\hat{u} \cdot \mathbf{x}_0)(1 + \cdots) \qquad \nabla\phi = ik\frac{e^{ikr}}{4\pi r}$$

$$\exp(-ik\hat{u} \cdot \mathbf{x}_0)(\hat{u} + \cdots)$$

where the indicated series are in descending powers of r and the convergence is uniform in $\hat{u}$. Furthermore, the surface element is

$$dS_r(\mathbf{x}) = r^2\,dS(\hat{u})$$

where $dS(\hat{u})$ is the surface element on the unit sphere. From the radiation condition, the limit of the integral over S_r is zero as $r \to \infty$; therefore $\mathbf{E}(\mathbf{x}_0) = 0$.

Similarly,

$$\mathbf{H}(\mathbf{x}_0) = \int_{S_r} [-i\omega\epsilon(\mathbf{E} \times \hat{u})\phi - (\hat{u} \times \mathbf{H}) \times \nabla\phi - (\hat{u} \cdot \mathbf{H})\,\nabla\phi]\,dS_r(\mathbf{x})$$

and from the radiation condition, $\mathbf{H}(\mathbf{x}_0) = 0$.

The unique field (4) which satisfies the radiation condition and is a solution of (3) is called the *radiation field generated by* $\mathbf{J}$, and $\mathbf{J}$ is called the *source* of the field. An arbitrary region R is called *source-free* if $\mathbf{J} = 0$ for all $\mathbf{x}$ in R.

We turn our attention now to the electromagnetic field in source-free regions, that is, solutions of (3) when $\mathbf{J} = 0$. First, let R be a finite region bounded by a surface S, with continuous unit normal $\hat{n}$. In contrast with the preceding case, we shall choose the normal directed *into* R. With this change, the argument above gives us, for $\mathbf{x}_0$ in R,

$$\mathbf{E}(\mathbf{x}_0) = \int_{S} [i\omega\mu(\hat{n} \times \mathbf{H})\phi + (\hat{n} \times \mathbf{E}) \times \nabla\phi + (\hat{n} \cdot \mathbf{E})\,\nabla\phi]\,dS(\mathbf{x})$$

$$\mathbf{H}(\mathbf{x}_0) = \int_{S} [-i\omega\epsilon(\hat{n} \times \mathbf{E})\phi + (\hat{n} \times \mathbf{H}) \times \nabla\phi + (\hat{n} \cdot \mathbf{H})\,\nabla\phi]\,dS(\mathbf{x})$$

Let $\mathbf{\nabla}_0$ be the gradient operator with respect to the point $\mathbf{x}_0$. Since

$$\mathbf{\nabla}_0 \times \mathbf{E}(\mathbf{x}_0) = i\omega\mu\mathbf{H}(\mathbf{x}_0)$$
$$\mathbf{\nabla} \times \mathbf{H}(\mathbf{x}_0) = -i\omega\epsilon\mathbf{E}(\mathbf{x}_0)$$
$$\mathbf{\nabla}_0\phi = -\mathbf{\nabla}\phi$$

it follows that

$$\int_S (\hat{n} \cdot \mathbf{E}) \,\mathbf{\nabla}\phi \, dS = -\frac{1}{i\omega\epsilon} \mathbf{\nabla}_0 \left[\mathbf{\nabla}_0 \cdot \int_S (\hat{n} \times \mathbf{H})\phi \, dS(\mathbf{x}) \right]$$

$$\int_S (\hat{n} \cdot \mathbf{H}) \,\mathbf{\nabla}\phi \, dS = \frac{1}{i\omega\mu} \mathbf{\nabla}_0 \left[\mathbf{\nabla}_0 \cdot \int_S (\hat{n} \times \mathbf{E})\phi \, dS(\mathbf{x}) \right]$$

Consequently,

$$\mathbf{E}(\mathbf{x}_0) = \mathbf{\nabla}_0 \times \int_S (\hat{n} \times \mathbf{E})\phi \, dS(\mathbf{x})$$

$$- \frac{1}{i\omega\epsilon} \mathbf{\nabla}_0(\mathbf{\nabla}_0 \cdot + k^2) \int_S (\hat{n} \times \mathbf{H})\phi \, dS(\mathbf{x})$$

$$\mathbf{H}(\mathbf{x}_0) = \mathbf{\nabla}_0 \times \int_S (\hat{n} \times \mathbf{H})\phi \, dS(\mathbf{x}) \tag{7}$$

$$+ \frac{1}{i\omega\mu} \mathbf{\nabla}_0(\mathbf{\nabla}_0 \cdot + k^2) \int_S (\hat{n} \times \mathbf{E})\phi \, dS(\mathbf{x})$$

It follows, then, that if R is a source-free region, finite and bounded by a surface S with continuous unit normal $\hat{n}$, the values of the electromagnetic field in R are completely determined by the values of $\hat{n} \times \mathbf{E}$ and $\hat{n} \times \mathbf{H}$ on S.

Now let us suppose that R is source-free but is the region exterior to a finite surface S, having a continuous unit normal. Let S_r be the surface of the sphere with center at $\mathbf{x} = 0$ and radius r, and let $r > |\mathbf{x}|$ for all $\mathbf{x}$ on S. Then if $\mathbf{x}_0$ is in R, and $|\mathbf{x}_0| < r$, the field at $\mathbf{x}_0$ has a representation differing from that in (7) only in that to each of the integrals over S must be added a similar integral over S_r. If the field satisfies the radiation conditions, then the same argument that was used to prove the uniqueness of the radiation field shows that as $r \to \infty$, the contribution of the integrals over S_r approaches zero. Then, if R is a source-free region which is the exterior of a finite surface S, having continuous unit normal vector $\hat{n}$, an electromagnetic field in R which satisfies the radiation conditions can be represented by (7).

The same type of calculation used in obtaining equations (5) shows the converse, that if the field in R is represented by (7), it must satisfy the radiation conditions.

CONDUCTING MATERIALS

A homogeneous isotropic region is said to be *conducting* if the relation $\mathbf{J} = \sigma\mathbf{E}$, where σ is a positive constant, holds. The value of σ is called the *conductivity* of the region. In a conducting region, $\mathbf{J}$ is removed from the second equation in (3) by replacing ϵ by $\epsilon + i\sigma/\omega$. Then the arguments that have just been used in obtaining representations (7) for the field in a source-free region continue to apply if k is replaced by $\omega\sqrt{\mu(\epsilon + i\sigma/\omega)}$ and the value of the square root whose real part (and consequently, the imaginary part) is taken to be positive.

From representation (7), we see that the presence of the factor $e^{ik|\mathbf{x}_0-\mathbf{x}|}$ in the integrand of the representation integrals has two related consequences. The first is that for fixed σ the field attenuates in an exponentially dominated fashion as the minimum distance between $\mathbf{x}_0$ and S increases; this phenomenon is known as the *skin effect*. The second is that if $\hat{n} \times \mathbf{E}$ and $\hat{n} \times \mathbf{H}$ remain bounded, independently of σ, then for every $\mathbf{x}_0$ in R the fields approach zero as $\sigma \to \infty$. It is useful to consider the idealization of a *perfectly conducting region*. For such a region, the fields are assumed to be zero at every interior point.

In applying conditions (2), special care must be taken when one of the two adjacent regions is regarded to be infinitely conducting. As mentioned before, conditions (2) are equivalent to the assumption of the integral form, obtained by using Stokes' theorem, of equations (1). The proof of this equivalence is obtained by choosing a small area A, having a point $\mathbf{x}$ of the common boundary S in its interior, lying on a plane which contains the normal vector to $\mathbf{x}$, bounded by a simple closed curve Γ, and equating

$$\oint_\Gamma \mathbf{E} \cdot \boldsymbol{\tau} \, d\Gamma = -\int_A \frac{\partial \mathbf{B}}{\partial t} \cdot \hat{n} \, dA$$

$$\int_\Gamma \mathbf{H} \cdot \boldsymbol{\tau} \, d\Gamma' = \int_A \left(\frac{\partial \mathbf{D}}{\partial t} + \mathbf{J}\right) \cdot \hat{n} \, dA$$

where $\boldsymbol{\tau}$ is the unit tangent vector on Γ and $\hat{n}$ is the correctly oriented normal vector on A. The proof of (2) then follows by allowing A to contrast and utilizing the continuity of $\mathbf{E}$ and $\mathbf{H}$ in each of the two regions. However, the proof depends upon the integrands of the surface integrals being bounded. In the first equation, this is the case even as σ becomes large. Hence if one of the regions is infinitely conducting, then $\hat{n} \times \mathbf{E} = 0$ at the surface of the other. However, in the second equation an indeterminacy is introduced as σ approaches infinity, so that $\hat{n} \times \mathbf{H}$ at the surface of the second region may have a nonzero limit value, which is interpreted as a *surface* current density.

We may now state the following *uniqueness theorem*.

Theorem 1 Let $\mathbf{E}$ and $\mathbf{H}$ be a harmonic electromagnetic field in a homogeneous isotropic source-free region R, which is the exterior of a finite surface S having a continuous normal vector $\hat{n}$.   If

$$\int_S (\mathbf{E} \times \tilde{\mathbf{H}} + \tilde{\mathbf{E}} \times \mathbf{H}) \cdot \hat{n} \, dS = 0$$

and if the radiation conditions (6) are satisfied, then $\mathbf{E}$ and $\mathbf{H}$ are identically zero for $|\mathbf{x}| > r_0$, where S is interior to $|\mathbf{x}| \le r_0$.

Proof

$$\boldsymbol{\nabla} \cdot (\mathbf{E} \times \tilde{\mathbf{H}} + \tilde{\mathbf{E}} \times \mathbf{H})$$
$$= (\boldsymbol{\nabla} \times \tilde{\mathbf{H}}) \cdot \mathbf{E} - (\boldsymbol{\nabla} \times \mathbf{E}) \cdot \tilde{\mathbf{H}} + (\boldsymbol{\nabla} \times \mathbf{H}) \cdot \tilde{\mathbf{E}} - (\boldsymbol{\nabla} \times \tilde{\mathbf{E}}) \cdot \mathbf{H} = 0$$

since

$$\boldsymbol{\nabla} \times \mathbf{E} = i\omega\mu\mathbf{H}$$
$$\boldsymbol{\nabla} \times \mathbf{H} = -i\omega\epsilon\mathbf{E}$$
$$\tilde{\mu} = \mu \qquad \tilde{\epsilon} = \epsilon$$

If, for all $\mathbf{x}$ on S, $|\mathbf{x}| < r$, then from Green's theorem,

$$\int_{S_r} [(\hat{n} \times \mathbf{E}) \cdot \tilde{\mathbf{H}} - (\hat{n} \times \mathbf{H}) \cdot \tilde{\mathbf{E}}] \, dS_r$$
$$= \int_{S_r} \hat{n} \cdot (\mathbf{E} \times \tilde{\mathbf{H}} + \tilde{\mathbf{E}} \times \mathbf{H}) \, dS_r = \int_S \hat{n} \cdot (\mathbf{E} \times \tilde{\mathbf{H}} + \tilde{\mathbf{E}} \times \mathbf{H}) \, dS = 0$$

In the first integral we may write

$$\hat{n} \times \mathbf{E} = \left(\hat{n} \times \mathbf{E} - \sqrt{\frac{\mu}{\epsilon}} \mathbf{H} \right) + \sqrt{\frac{\mu}{\epsilon}} \mathbf{H}$$
$$\hat{n} \times \mathbf{H} = \left(\hat{n} \times \mathbf{H} + \sqrt{\frac{\epsilon}{\mu}} \mathbf{E} \right) - \sqrt{\frac{\epsilon}{\mu}} \mathbf{E}$$

Then, from the radiation condition,

$$\lim_{r \to \infty} \int_{S_r} \left(\sqrt{\frac{\mu}{\epsilon}} |H|^2 + \sqrt{\frac{\epsilon}{\mu}} |E|^2 \right) dS_r = 0$$

Consequently, if S is interior to the surface $|\mathbf{x}| = r_0$ and R_r is the region $r_0 < |\mathbf{x}| < r$,

$$\lim_{r \to \infty} \frac{1}{r} \int_{R_r} \left(\sqrt{\frac{\mu}{\epsilon}} |H|^2 + \sqrt{\frac{\epsilon}{\mu}} |E|^2 \right) dR_r = 0$$

We shall now show that if $|E|$ and $|H|$ are not identically zero for $\mathbf{x}$ in R, then there exists a constant $C > 0$ such that if r is sufficiently large,

$$\int_{R_r} \left(\sqrt{\frac{\mu}{\epsilon}} |H|^2 + \sqrt{\frac{\epsilon}{\mu}} |E|^2 \right) dR_r \ge Cr$$

This will provide the contradiction that proves the theorem.   Since the rectangular components of $\mathbf{E}$ and $\mathbf{H}$ satisfy the scalar equation

$$\nabla^2 f + k^2 f = 0$$

in R, it is sufficient to prove that if f is not identically zero for $|x| \geq r_0$ there exists a $C > 0$ such that for r sufficiently large

$$\int_{R_r} |f|^2 \, dR_r \geq Cr$$

The proof of this is called *Rellich's theorem*. If f is not identically zero for $r \geq r_0$, and if the numbers

$$\int_0^{2\pi} \int_0^{\pi} |f|^2 P_n^{|m|}(\cos\theta) e^{-im\phi} \sin\theta \, d\theta \, d\psi \qquad n = 0, 1, 2, \ldots,$$
$$-n \leq m \leq n$$

are not all identically zero for $r \geq r_0$ and g is one of these numbers, then, from Bessel's inequality,

$$\int_0^{2\pi} \int_0^{\pi} |f|^2 \sin\theta \, d\theta \, d\psi \geq \frac{2n+1}{4\pi} \frac{(n-|m|)!}{(n+|m|)!} |g|^2$$

(see the problems at the end of Chap. 4).

It is now sufficient to show that there exists a constant $C_1 > 0$ such that for r sufficiently large,

$$\int_{r_0}^{r} r^2 |g|^2 \, dr \geq C_1 r$$

Now, if $h = rg$,

$$\frac{d^2 h}{dr^2} + k^2 h = \frac{n(n+1)}{r^2} h$$

From the methods of Chap. 8, h must be of the form

$$h = A_1 u_1 + A_2 u_2$$

where A_1 and A_2 are constants, $|A_1|^2 + |A_2|^2 \neq 0$, and the u_j are the linearly independent solutions of the integral equation

$$u_j = e^{\pm ikr} + \frac{n(n+1)}{k} \int_r^{\infty} \sin k \, (\zeta - r) u_j(\zeta) \frac{d\zeta}{\zeta^2} \qquad j = 1, 2$$

The positive and negative signs in the exponential correspond, respectively, to $j = 1, 2$. Then, if $w_j = u_j e^{\mp ikr}$, the functions $r(w_j - 1)$ are bounded as $r \to \infty$. Consequently,

$$\int_{r_0}^{r} |h|^2 \, dr = (|A_1|^2 + |A_2|^2)(r - r_0) + D$$

where

$$|D| \leq D_1 + D_2 \log \frac{r}{r_0}$$

D_1 and D_2 being constants.   Therefore, if r is sufficiently large,

$$\frac{1}{r} \int_{r_0}^{r} |h|^2 \, dr \geq \frac{1}{2} \left(|A_1|^2 + |A_2|^2 \right)$$

which completes the proof.

BOUNDARY-VALUE PROBLEMS OF ELECTROMAGNETIC THEORY

Diffraction   The general problem of diffraction consists of the following elements.   A region R is given as the exterior of a finite surface S_0, which may have several closed components. R is usually taken to be free space. A harmonic electromagnetic field $(\mathbf{E}^P, \mathbf{H}^P)$ is given in R, satisfying the source-free Maxwell's equations.   This field is called the *primary*, or *source, field*.   The total electromagnetic field $(\mathbf{E}, \mathbf{H})$ is assumed to be of the form

$$(\mathbf{E} = \mathbf{E}^P + \mathbf{E}^D, \mathbf{H} = \mathbf{H}^P + \mathbf{H}^D)$$

where $(\mathbf{E}^D, \mathbf{H}^D)$ is called the *secondary*, or *diffracted, field*.   The diffracted field is required to satisfy the radiation condition.   The problem is to find the refracted field.

If S has a continuous normal vector, and the interior of S is homogeneous and isotropic (but not conducting), the problem is said to be *well-set*. Specifically, in the region $|\mathbf{x}| \geq r_0$ containing no points of S, there is at most one diffracted field which satisfies the given conditions.   The linearity of Maxwell's equations shows that this statement is equivalent to the statement that if the primary field is zero, the diffracted field must be zero for $|\mathbf{x}| \geq r_0$.   This will follow, from the preceding section, if

$$\int_{S} \hat{n} \cdot (\mathbf{E} \times \tilde{\mathbf{H}} + \tilde{\mathbf{E}} \times \mathbf{H}) \, dS = 0$$

Now, this integral depends only on the tangential components of $\mathbf{E}$ and $\mathbf{H}$ in R at S, and according to conditions (2), it does not change when $\mathbf{E}$ and $\mathbf{H}$ are the fields interior to S.   But then, from Green's theorem, this integral can be replaced by an integral, with the same integrand, over an arbitrarily small surface, and it is therefore equal to zero.

If the region interior to S is conducting, with conductivity σ, the integral is not zero but is equal to

$$-\sigma \int_{R'} |E|^2 \, dR'(\mathbf{x})$$

integrated over R', the interior of S.   But, from the proof of the uniqueness theorem, this gives

$$\lim_{r \to 0} \int_{S_r} \left(\sqrt{\frac{\mu}{\epsilon}} |H|^2 + \sqrt{\frac{\epsilon}{\mu}} |E|^2 \right) dS_r = -\sigma \int_{R_1} |E|^2 \, dR'(\mathbf{x})$$

Since the integral over S_r is nonnegative, the integral on the right must be zero, and the proof of uniqueness proceeds as before, with the additional information that the field interior to S must be zero.

If the interior of S is perfectly conducting, $\hat{n} \times \mathbf{E} = 0$ on S, or ij, the field satisfies an *impedance boundary condition* on S,

$$\hat{n} \times \mathbf{E} = -i\eta\hat{n} \times (\hat{n} \times \mathbf{H})$$

where η is a real constant, and the integral is again zero. Furthermore, if the interior of S consists of several homogeneous isotropic regions contained within each other, with common surfaces which have continuous normal vectors, the uniqueness argument can be extended. Thus for a wide variety of configurations the requirement that the diffracted field satisfy the radiation condition assures that at most one solution exists to the mathematical problem. That this solution is in fact the meaningful physical solution follows from the physical assumption that since the diffracted field must arise through the creation by the primary field of motion of charge in the material of which S consists, the diffracted field is a radiation field generated by a current which describes this motion of charge.

In the case where S does not have a continuous normal vector, the divergence theorem cannot be used to move the integrand of

$$\int_S \hat{n} \cdot (\mathbf{E} \times \tilde{\mathbf{H}} + \tilde{\mathbf{E}} \times \mathbf{H})\, dS$$

out to a surface S_r, which is, after all, the crux of the uniqueness theorem. In this case, to assure uniqueness, conditions are placed on the electromagnetic field which permit S to be replaced by an adjacent surface which does have a continuous normal, and on which the integrand being zero is equivalent to its being zero on S. Conditions of this type are called *edge conditions* and have various physical interpretations whose justification lies in the experimental verification of the prediction.

If S is not finite, the uniqueness theorem is no longer valid. However, if the idealization is made that S is the surface of a cylinder whose generatrix is a simple closed curve and whose plane is perpendicular to the directrix, and if all fields are assumed to be independent of displacement in the direction of the directrix, the mathematical problem becomes one in the two-dimensional euclidean space, and results which are simpler, but similar to those already obtained, are valid.

Now that we have stated the diffraction problem and have shown that for a large class of configurations the unique solution to the mathematical problem has physical meaning, let us turn to the question of finding the solution. If the components of the electromagnetic field can be represented as an infinite series (convergent in some generalized sense) of well-defined functions, or as an integral whose integrand consists of well-defined functions, the solution is said to be in *closed form*. The classical method of finding closed-form solutions is to utilize, in various ways, solutions of the equation

$$\nabla^2 \psi + k^2 \psi = 0$$

For example, in the problem of diffraction by a sphere, if we allow ψ to have the form

$$\psi = f(r)e^{im\phi}P_n^{|m|}(\cos \theta)$$

where $f(r)$ is a solution of

$$\frac{d}{dr}\left(r^2 \frac{df}{dr}\right) + [k^2r^2 - n(n+1)]f = 0$$

the fields are represented as series, with constant coefficients, of vector-valued functions of the form

$$\mathbf{M} = \nabla\psi \times \mathbf{x} \qquad \text{and} \qquad \mathbf{N} = \nabla \times \mathbf{M}$$

Using the known properties of the functions

$$e^{im\phi}P_n^{|m|}(\cos \theta)$$

we can show that if $\mathbf{A}_1$ and $\mathbf{A}_2$ are any two different functions of the class of functions of this type, where the difference is caused either by one being of type $\mathbf{M}$ and the other of type $\mathbf{N}$ or by a difference between m and n, then

$$\int_0^{2\pi} \int_0^{\pi} \mathbf{A}_1 \cdot \tilde{\mathbf{A}}_2 \sin \theta \, d\theta \, d\phi = 0$$

Thus these functions form a type of orthogonal system, and this property is used to determine the unknown coefficients in the expansion. There are only a finite number of geometric configurations for which this method can be used; approximation methods for other configurations are not usually within the framework of the material covered in this text.

Propagation in a fixed direction   In three-dimensional euclidean space, let $\hat{\xi}$ be a unit vector, and let $\mathbf{x} \cdot \hat{\xi} = \xi$.   We seek solutions of the source-free Maxwell's equations of the form

$$\mathbf{E}(\mathbf{x}) = \mathbf{A}e^{i\gamma\xi} \qquad \mathbf{H}(\mathbf{x}) = \mathbf{B}e^{i\gamma\xi}$$

where $\mathbf{A}$ and $\mathbf{B}$ are independent of ξ and γ is a constant.   If

$$\mathbf{A} = \mathbf{A}_T + \phi\hat{\xi} \qquad \mathbf{B} = \mathbf{B}_T + \psi\hat{\xi}$$

where $\mathbf{A}_T$ and $\mathbf{B}_T$, the *transverse* components of $\mathbf{A}$ and $\mathbf{B}$, are perpendicular to $\hat{\xi}$, and if

$$\mathbf{\nabla}_T = \mathbf{\nabla} - \hat{\xi}\frac{\partial}{\partial\xi} = \mathbf{\nabla} - i\gamma\hat{\xi}$$

the following equations are satisfied:

$$i\gamma(\hat{\xi} \times \mathbf{A}_T) - i\omega\mu\mathbf{B}_T = \hat{\xi} \times \mathbf{\nabla}_T\phi \qquad \mathbf{\nabla}_T \cdot (\hat{\xi} \times \mathbf{A}_T) = -i\omega\mu\psi$$
$$i\omega\epsilon\mathbf{A}_T + i\gamma(\hat{\xi} \times \mathbf{B}_T) = \hat{\xi} \times \mathbf{\nabla}_T\psi \qquad \mathbf{\nabla}_T \cdot (\hat{\xi} \times \mathbf{B}_T) = i\omega\epsilon\phi \qquad (8)$$

When $\psi = \phi = 0$, the field is said to be *transverse*. If only $\phi = 0$, the field is said to be *transverse electric*. If only $\psi = 0$, the field is said to be *transverse magnetic*.

If the field is transverse, from (8),

$$\gamma^2 = k^2$$

$$\hat{\xi} \times \mathbf{B}_T = \frac{\omega\epsilon}{\gamma}\mathbf{A}_T$$

$$\mathbf{\nabla}_T \times \mathbf{A}_T = 0 \qquad \mathbf{\nabla}_T \cdot \mathbf{A}_T = 0$$

Consequently,

$$\mathbf{A}_T = \mathbf{\nabla}_T f$$

where f is a function for which $\nabla_T^2 f = 0$, a solution of the two-dimensional Laplace equation. If the field is to be bounded in the entire three-dimensional space, $\mathbf{\nabla}_T f$ must be a constant vector, orthogonal, of course, to $\hat{\xi}$. Electromagnetic fields of this type are called *plane waves*.

For transverse electric and transverse magnetic waves, ψ and ϕ satisfy the equation

$$\nabla_T^2 f + (k^2 - \gamma^2)f = 0$$

and in the respective cases,

$$\mathbf{A}_T = \frac{i\omega\mu\hat{\xi} \times \mathbf{\nabla}\psi}{k^2 - \gamma^2} \qquad \mathbf{B}_T = \frac{i\gamma\,\mathbf{\nabla}\psi}{k^2 - \gamma^2}$$

and

$$\mathbf{A}_T = \frac{i\gamma\,\mathbf{\nabla}\phi}{k^2 - \gamma^2} \qquad \mathbf{B}_T = -\frac{i\omega\epsilon\hat{\xi} \times \mathbf{\nabla}\phi}{k^2 - \gamma^2}$$

WAVEGUIDES

Let the region R be the interior of a cylinder whose directrix is parallel to the vector $\hat{\xi}$ and whose generatrix is the boundary of a finite area orthogonal to $\hat{\xi}$. Let the region exterior to R be thought of as a perfect conductor, so that the tangential component of $\mathbf{A}$ is zero at the boundary of R. Such a region is called a *waveguide*, and the types of harmonic fields which can exist in such a structure are of great importance. In a wave-guide, these fields are specified by the boundary conditions on f, for the three types of waves considered. If Γ is the boundary of the generatrix, and if Γ is piecewise differentiable, these conditions are, respectively, that f is constant on each of the differentiable pieces of Γ, $\partial f / \partial n = 0$ on Γ, and $f = 0$ on Γ. For the transverse electromagnetic case, the boundary condition and the equation $\nabla_T{}^2 f = 0$ determine all fields. For the other cases, the boundary conditions and the equation

$$\nabla_T{}^2 f + (k^2 - \gamma^2)f = 0$$

determine a countable number of values of γ, and for each of them a countable number of fields of the specified type.

CAVITIES

A generalization of the preceding case is that in which R is the interior of a closed and bounded surface. Source-free electromagnetic fields are sought in R which satisfy the boundary condition $\hat{n} \times \mathbf{E} = 0$ on S. Again it can be shown that only a countable number of such fields exist, and that there are, in fact, only a countable number of values of k for which they can exist. In both the waveguide and the cavity problem, for only a certain finite number of configurations can the solution be given in closed form.

appendix C

The Heat Equation

HEAT TRANSFER

In the theory of continuum dynamics developed in Appendix A, the thermodynamic processes were assumed to be either isothermal or adiabatic. At the other end of the scale, we can assume that all changes of energy density are accounted for by heat transfer, with any changes caused by deformation or dissipation considered negligible. In this case, we account for these changes by equating the volume integral

$$C_V \int_R \frac{\partial \theta(\mathbf{x},t)}{\partial t} \, dR(\mathbf{x})$$

where

R = an arbitrary region
C_V = a positive constant, the specific heat at constant volume
$\theta(\mathbf{x},t)$ = temperature

to the surface integral

$$\int_S \mathbf{Q}(\mathbf{x},t) \cdot \hat{n}(\mathbf{x}) \, dS(\mathbf{x})$$

where

S = the boundary of R
$\hat{n}$ = the unit normal outward on S
$\mathbf{Q}(\mathbf{x},t)$ = the heat-flux vector

Since R is arbitrary,

$$C_V \frac{\partial \theta}{\partial t} + \mathbf{\nabla} \cdot \mathbf{Q} = 0$$

If we further assume that

$$\mathbf{Q}(\mathbf{x},t) = -\lambda \, \mathbf{\nabla} \theta(\mathbf{x},t)$$

where λ, a positive constant, is the thermal conductivity, we obtain the *heat equation*,

$$\frac{\partial \theta}{\partial t} = k^2 \, \nabla^2 \theta$$

where $k^2 = \lambda/C_V$.

BOUNDED REGIONS

Suppose that the continuum has a closed surface S as a boundary. The physical condition that the boundary is maintained at constant temperature is expressed as θ = constant on S. The physical condition that the boundary is insulated is expressed as $\mathbf{Q} \cdot \hat{n} = 0$ on S, where $\hat{n}$ is the unit normal on S.

Some general remarks can be made about the solution of the heat equation in these two circumstances.

θ **constant on** S Because adding a constant to θ does not affect the status of θ as a solution of the heat equation, no generality is lost in assuming that $\theta = 0$ on S. In a fashion analogous to the Sturm-Liouville problem in Chap. 8, we can study the equation

$$-\nabla^2 u = \lambda u$$

where λ is a constant and u not identically zero is required to be zero on S. Setting

$$(f,g) = \int f(\mathbf{x})g(\mathbf{x}) \, dR(\mathbf{x})$$

where the integral is extended over the interior of S, we have

$$\lambda(u,u) = -(\nabla^2 u,u) = (\nabla u, \nabla u) \geq 0$$

If $(\nabla u, \nabla u) = 0$ and $u = 0$ on S, then u is identically zero; thus $\lambda > 0$. As in the fundamental theorem of the Sturm-Liouville problem, there exists an increasing sequence of positive numbers $\lambda_1, \lambda_2, \ldots$ and functions $u_1, u_2, \ldots$ such that

$$-\nabla^2 u_n = \lambda_n u_n \qquad (u_n, u_m) = \delta_{nm}$$

and for any function u which is zero on S and which has continuous second derivatives,

$$\lim_{N \to \infty} \int \left| u - \sum_{n=1}^{N} (u,u_n)u_n \right|^2 dR(\mathbf{x}) = 0$$

If $\theta(\mathbf{x},t)$ is a solution of the heat equation having continuous second derivatives (with respect to $\mathbf{x}$), and if $\theta = 0$ on S,

$$\frac{\partial(\theta,u_n)}{\partial t} = k^2(\nabla^2\theta,u_n) = -\lambda_n k^2(\theta,u_n)$$

Thus

$$(\theta,u_n) = a_n e^{-\lambda_n k^2 t}$$

where a_n is a constant.

If $\lim_{t \to 0} \theta(\mathbf{x},t)$ is given to be $\theta_0(\mathbf{x})$, then

$$(\theta,u_n) = (\theta_0,u_n)e^{-\lambda_n k^2 t}$$

The function $\theta_0(\mathbf{x})$ is called the *initial temperature distribution*. For any two solutions θ' and θ'' having the same initial distribution,

$$(\theta',u_n) = (\theta'',u_n) \qquad \text{for all } n$$

and therefore

$$\int |\theta' - \theta''|^2 \, dR(\mathbf{x}) = 0$$

which implies, for continuum functions, that θ' and θ'' are identical. Thus if there is a solution, it is completely determined by the initial conditions. Furthermore, since $\lambda_n > 0$ for all n,

$$\lim_{t \to \infty} (\theta,u_n) = 0$$

When this happens we say that θ *converges weakly* to zero; this is called the *equilibrium state* of the continuum. The equilibrium state is independent of the initial condition.

Whether a solution exists for the mathematical problem as stated is an open question. In many problems it is easy to show that the infinite series

$$\sum_{n=1}^{\infty} (\theta_0, u_n) e^{-\lambda_n k^2 t} u_n(\mathbf{x}) \qquad \text{for } t > 0$$

converges uniformly to a function $\theta(\mathbf{x}, t)$ of the heat equation, which satisfies the boundary condition. Furthermore,

$$\lim_{t \to 0} (\theta, u_n) = (\theta_0, u_n)$$

and so θ converges weakly to θ_0 as $t \to 0$.

$\boldsymbol{\nabla}\theta \cdot \hat{n} = 0$ **on** S The argument is almost identical to that in the preceding case, except that λ_1, the smallest value of the sequence $\lambda_1, \lambda_2, \ldots$, is zero, and the corresponding u is a constant $\sqrt{1/V}$, where V is the volume of the interior of S. For a solution $\theta(\mathbf{x}, t)$,

$$(\theta, u_n) = (\theta_0, u_n) e^{-\lambda_n k^2 t}$$

and the equilibrium distribution, to which θ converges weakly as $t \to \infty$, is the constant

$$\frac{1}{V} \int \theta_0(\mathbf{x}) \, dR(\mathbf{x})$$

THE GENERAL CASE OF EQUILIBRIUM

Instead of considering the equilibrium state as the limit, as $t \to \infty$, of the solution $\theta(\mathbf{x}, t)$, let us define the general case of the equilibrium problem as the case in which $\partial\theta/\partial t = 0$. Then the heat equation becomes the Laplace equation,

$$\nabla^2 \theta(\mathbf{x}) = 0$$

and solutions can be sought for which either θ or $\boldsymbol{\nabla}\theta \cdot \hat{n}$ are specified as functions on the boundary.

THE UNBOUNDED CASE

In this case we assume that the continuum is not bounded. The simplest situation is that in which we assume that the continuum is the entire space and that for each t, $\theta = 0$ for $|\mathbf{x}|$ sufficiently large. Then

$$\hat{\theta}(\boldsymbol{\omega}, t) = \int \theta(\mathbf{x}, t) \exp(-i\boldsymbol{\omega} \cdot \mathbf{x}) \, dR(\mathbf{x})$$

where $\boldsymbol{\omega}$ is a three-dimensional vector and $\hat{\theta}(\boldsymbol{\omega},t)$ is the *spatial Fourier transform* of θ.   As in the preceding cases,

$$\frac{d\hat{\theta}}{dt} = -k^2|\boldsymbol{\omega}|^2\hat{\theta}$$

so that

$$\hat{\theta} = a_\omega \exp\left(-k^2|\boldsymbol{\omega}|^2 t\right)$$

where $a_\omega = \hat{\theta}_0(\boldsymbol{\omega})$, the transform of the initial distribution.   As before, the solution is determined by $\theta_0(\mathbf{x})$, and if the Fourier integral

$$\frac{1}{(2\pi)^3}\int \hat{\theta}_0(\boldsymbol{\omega}) \exp\left(-k^2|\boldsymbol{\omega}|^2 t\right) \exp\left(i\boldsymbol{\omega}\cdot\mathbf{x}\right) dV(\boldsymbol{\omega})$$

converges for all t, it converges to the solution.

Index